# Computer Applications
in the Polymer Laboratory

ACS SYMPOSIUM SERIES **313**

# Computer Applications in the Polymer Laboratory

**Theodore Provder,** EDITOR
*Glidden Coatings and Resins*

Developed from a symposium sponsored by
the Division of Polymeric Materials Science and Engineering
at the 189th Meeting
of the American Chemical Society,
April 28–May 3, 1985

American Chemical Society, Washington, DC 1986

**Library of Congress Cataloging-in-Publication Data**

Computer applications in the polymer laboratory.
(ACS symposium series, ISSN 0097-6156; 313)

Includes bibliographies and indexes.

1. Polymers and polymerization—Data processing—Congresses.

I. Provder, Theodore, 1939–   . II. American Chemical Society. Division of Polymeric Materials Science and Engineering. III. American Chemical Society (189th: 1985: Miami Beach, Fla.) IV. Series.

QD381.9.E4C66    1986    668.9′028′5    86-10831
ISBN 0-8412-0977-4

D
668.4028'54
com

Copyright © 1986

American Chemical Society

All Rights Reserved. The appearance of the code at the bottom of the first page of each chapter in this volume indicates the copyright owner's consent that reprographic copies of the chapter may be made for personal or internal use or for the personal or internal use of specific clients. This consent is given on the condition, however, that the copier pay the stated per copy fee through the Copyright Clearance Center, Inc., 27 Congress Street, Salem, MA 01970, for copying beyond that permitted by Sections 107 or 108 of the U.S. Copyright Law. This consent does not extend to copying or transmission by any means—graphic or electronic—for any other purpose, such as for general distribution, for advertising or promotional purposes, for creating a new collective work, for resale, or for information storage and retrieval systems. The copying fee for each chapter is indicated in the code at the bottom of the first page of the chapter.

The citation of trade names and/or names of manufacturers in this publication is not to be construed as an endorsement or as approval by ACS of the commercial products or services referenced herein; nor should the mere reference herein to any drawing, specification, chemical process, or other data be regarded as a license or as a conveyance of any right or permission, to the holder, reader, or any other person or corporation, to manufacture, reproduce, use, or sell any patented invention or copyrighted work that may in any way be related thereto. Registered names, trademarks, etc., used in this publication, even without specific indication thereof, are not to be considered unprotected by law.

PRINTED IN THE UNITED STATES OF AMERICA

# ACS Symposium Series

## M. Joan Comstock, *Series Editor*

### *Advisory Board*

Harvey W. Blanch
University of California—Berkeley

Alan Elzerman
Clemson University

John W. Finley
Nabisco Brands, Inc.

Marye Anne Fox
The University of Texas—Austin

Martin L. Gorbaty
Exxon Research and Engineering Co.

Roland F. Hirsch
U.S. Department of Energy

Rudolph J. Marcus
Consultant, Computers &
 Chemistry Research

Vincent D. McGinniss
Battelle Columbus Laboratories

Donald E. Moreland
USDA, Agricultural Research Service

W. H. Norton
J. T. Baker Chemical Company

James C. Randall
Exxon Chemical Company

W. D. Shults
Oak Ridge National Laboratory

Geoffrey K. Smith
Rohm & Haas Co.

Charles S. Tuesday
General Motors Research Laboratory

Douglas B. Walters
National Institute of
 Environmental Health

C. Grant Willson
IBM Research Department

# FOREWORD

The ACS SYMPOSIUM SERIES was founded in 1974 to provide a medium for publishing symposia quickly in book form. The format of the Series parallels that of the continuing ADVANCES IN CHEMISTRY SERIES except that, in order to save time, the papers are not typeset but are reproduced as they are submitted by the authors in camera-ready form. Papers are reviewed under the supervision of the Editors with the assistance of the Series Advisory Board and are selected to maintain the integrity of the symposia; however, verbatim reproductions of previously published papers are not accepted. Both reviews and reports of research are acceptable, because symposia may embrace both types of presentation.

# CONTENTS

Preface ............................................................................. ix

### LABORATORY INFORMATION GENERATION, MANAGEMENT, AND ANALYSIS TOOLS

1. **Laboratory Automation: A New Perspective** ........................... 2
   Mark E. Koehler

2. **Economic Considerations of Laboratory Information Management Systems** ........................................................................ 6
   Joseph H. Golden

3. **Applications of Computer Data Base Management in Polymer and Coatings Research** ............................................. 17
   Mark E. Koehler, A. F. Kah, and T. F. Niemann

4. **Advances in Scientific Software Packages** ............................ 23
   Channing H. Russell

5. **Computer-Assisted Polymer Design** ..................................... 31
   Rudolph Potenzone, Jr., and David C. Doherty

6. **Silicone Acrylate Copolymers: Designed Experiment Success** ........... 39
   T. R. Williams and M. D. Nave

7. **Analysis and Optimization of Constrained Mixture-Design Formulations** .... 58
   Stephen E. Krampe

### INSTRUMENT AUTOMATION FOR POLYMER CHARACTERIZATION

8. **Advantages of Interfacing a Viscoelastic Device with a High-Speed and -Capacity Computer and an Advanced Statistical–Graphics Software Package** ...................................................................... 76
   Stephen Havriliak, Jr.

9. **Analysis of Isochronal Mechanical Relaxation Scans** .................. 89
   Richard H. Boyd

10. **Automated Rheology Laboratory: Part I** ............................. 105
    V. G. Constien, E. L. Fellin, M. T. King, and G. G. Graves

11. **Automated Rheology Laboratory: Part II** ............................ 114
    M. T. King, V. G. Constien, and E. L. Fellin

12. **An Automated Analysis System for a Tensile Tester** ................ 123
    T. T. Gill and Mark E. Koehler

13. **Software for Data Collection and Analysis from a Size-Exclusion Liquid Chromatograph** ............................... 130
    John D. Barnes, Brian Dickens, and Frank L. McCrackin

14. **An Automated Apparatus for X-ray Pole Figure Studies of Polymers** .... 140
    John D. Barnes and E. S. Clark

15. Computers and the Optical Microscope ............................ 155
    M. B. Rhodes and R. P. Nathhorst

    POLYMERIZATION AND CURE PROCESS MODELING
    AND CONTROL

16. Modeling and Simulation Activities in a Large Research and Development Laboratory for Coatings ............................................. 170
    D. T. Wu

17. Flexible Control of Laboratory Polymer Reactors by Using Table-Driven Software .................................... 179
    Robert Albrecht-Mallinger

18. Application of State Variable Techniques to the Control of a Polystyrene Reactor ........................................... 187
    David J. Hild, Richard E. Gilbert, and Delmar C. Timm

19. Initiation Reactions and the Modeling of Polymerization Kinetics ........ 202
    L. H. Garcia-Rubio and J. Mehta

20. Mathematical Modeling of Emulsion Polymerization Reactors: A Population Balance Approach and Its Applications .................. 219
    A. Penlidis, J. F. MacGregor, and A. E. Hamielec

21. Kinetics Analysis of Consecutive Reactions Using Nelder–Mead Simplex Optimization ..................................................... 241
    Gary M. Carlson and Theodore Provder

22. Development and Application of Network Structure Models to Optimization of Bake Conditions for Thermoset Coatings ............. 256
    David R. Bauer and Ray A. Dickie

23. A Kinetic Study of an Anhydride-Cured Epoxy Polymerization .......... 275
    C. C. Lai, Delmar C. Timm, B. W. Eaton, and M. D. Cloeter

24. Investigation of the Self-Condensation of 2,4-Dimethylol-*o*-cresol by ¹H-NMR Spectroscopy and Computer Simulation ................... 288
    Alexander P. Mgaya, H. James Harwood, and Anton Sebenik

INDEXES

Author Index ...................................................... 315

Subject Index ..................................................... 315

# PREFACE

IMPROVEMENTS IN COMPUTER EQUIPMENT and decreases in its cost have sparked the personal computer reovlution that has had a significant impact on all aspects of our society. This technological revolution has also affected the research and development (R&D) worker (scientist, technologist, manager, etc.) in the field of applied polymer science. The technology now allows the concept of task automation to become a reality. The present-day R&D scientist and technologist must function effectively on two sides of the laboratory: the bench side where instruments are located and experiments are performed and the desk side where data are analyzed, reports and publications are written, and desk and information management tasks are performed. Therefore, the R&D scientist and technologist need access to technical and office tools. The present-day R&D manager also has to function in a dual mode by bridging the technical and business arenas. The R&D manager needs access to many of the same technical and office tools used by the scientist and technologist but also requires the use of business information management and decision support tools. Computer technology is beginning to allow the R&D worker to have access to all of these tools through a universal work station coupled to a computer network, which facilitates transfer of information and communications within and between the technical and business functions.

The topics in this book are divided into three sections and reflect the impact of changes in computer technology during the last 5 years upon applied polymer science. The first section deals with laboratory information generation, management, and analysis tools, which are some of the elements required for task automation. The second section covers the field of instrument automation for polymer characterization. Instrument automation has been facilitated by the availability of inexpensive powerful microcomputers and peripherals as well as easier to use software. Laboratory scientists do not have to get down to the "bits and bytes" level of software development to the extent they previously had to. The use of robotics is aiding the automation of the total analysis, including sample handling, and is allowing automation of a series of analyses on the same sample. The current technology also makes it more cost-effective to automate one-of-a-kind instrumentation. The third section covers the field of polymerization and cure process modeling and control. The availability of very powerful laboratory micro-, supermicro-, and superminicomputers allows the laboratory scientist to be much more cost-effective in doing extensive fundamental modeling and simulation than was previously possible with large mainframe batch computers. This trend is expected to accelerate in the future.

It has been said that the past is prologue to the future. This statement is certainly true for the impact of improvements in computer technology during the last 5 years upon applied polymer science. Looking toward the future, one can speculate on the continuing impact of improvements on enhanced computer technology over the next 5 years. The existing trends will accelerate. Easy-to-use sophisticated scientific software packages including a wide variety of complex modeling and simulation will become more available. The need for the laboratory scientist to write software using a computer language will almost disappear. The growth in the use of robotics will increase at an almost exponential rate. The development of vast computerized technical libraries and data bases will only be limited by the effort required to input data.

A new technology that is expected to have a significant impact will be the application of expert systems to applied polymer science. One can envision expert systems such as (1) self-optimizing automated instrumental analyses, (2) molecular modeling for designer polymers having a specific chemical–physical property profile, (3) formulation applications (e.g., coatings, polymer composites, cosmetics, food preparations, etc.) in which cost and performance are optimized, and (4) diagnostic and prognostic applications in which fuzzy logic is applied to semiquantitative and qualitative information. The future impact of enhanced computer technology upon applied polymer science, indeed, appears to be very exciting.

I thank the authors for their effective oral and written communications and the reviewers for their critiques and constructive comments. I also acknowledge the book jacket design concepts provided by Ann F. Kah.

THEODORE PROVDER
Glidden Coatings and Resins
SCM Corporation
Strongsville, OH 44136

January 31, 1986

# LABORATORY INFORMATION GENERATION, MANAGEMENT, AND ANALYSIS TOOLS

# 1

# Laboratory Automation: A New Perspective

**Mark E. Koehler**

**Dwight P. Joyce Research Center, Glidden Coatings and Resins, SCM Corporation, Strongsville, OH 44136**

```
              Laboratory automation has traditionally meant
              laboratory instrument automation.  While the automated
              collection and analysis of data from laboratory
              instruments is still a significant part of laboratory
              automation, in the modern automated laboratory it is
              only a part of a larger perspective with the focus on
              task automation.  Simply stated, the goal should be to
              automate tasks, not instruments.
                  This expanded view of task automation includes new
              capabilities in the the traditional area of instrument
              automation and in the somewhat newer related field of
              robotics.  In addition it includes a number of functions
              which are not new to the office and business
              environment but have only recently become readily
              available in the laboratory.  These are tools such as
              data base management, scientific text processing, and
              electronic mail and document transfer.  One way to
              improve technical productivity is by giving the
              scientist more time to do science.  This can be
              accomplished through improved efficiency in the office,
              communication, and information retrieval functions
              which must be performed as well as by allowing science
              to be done in new and more efficient ways through the
              use of computers.
```

The explosive growth in the availability of computer tools in the laboratory requires a new look at the concept of laboratory automation. Much larger gains in the efficiency and effectiveness of research can be realized by automating tasks rather than by simply automating instruments. Research is done by researchers, not by instruments. Instruments are just one of many tools which can be used by the researcher. This paper will attempt to give an overview of instrument automation as the traditional view of laboratory automation, and extend this concept to the automation of the total task of research.

0097-6156/86/0313-0002$06.00/0
© 1986 American Chemical Society

## Instrument Automation

It is interesting to trace the development of instrument automation over the relatively brief period of the past ten to fifteen years. Early in this period, a truly automated instrument was a rare and expensive item built around a costly dedicated minicomputer. Automated data collection and analysis from any instrument which was not automated at the factory was usually accomplished by digitizing the data and storing it on a transportable media such as paper tape. These data were then delivered and fed to a timeshare system of some sort on which the data reduction program ran and which printed a report and sometimes a plot of the data. Often a considerable time delay occured between the generation and the analysis of the data. The scientist was at the mercy of the computer elite who could implement his data logger and provide the necessary computer resources to analyze his data. The process was expensive, both in time and in money.

With the advent of the inexpensive microprocessor chip a number of things began to happen. The laboratory scientist found that he could gain a new independence. Data could be collected and analyzed virtually simultaneously without having to carry his data to the tabernacle of the computer for analysis and without having to spending a significant portion of his research budget in the process. Instrument manufacturers began to implant one or more microprocessors into their wares. In some cases these instruments did absolutely nothing that their analog predecessors did not do, but they did it with style and sold new instruments. In the case of pre-automation instruments or the cases where the vendors did not supply what the experiment demanded, the determined scientist, with very little cash investment and with a great deal of effort, frequently applied in large doses late at night, could now do what he had always wanted. Home-brew automation was still beyond the capabilities and inclinations of the average scientist, but all the while, the instrument vendors were doing a better job of supplying what he needed anyway.

While all this was happening more powerful computers were getting cheaper, cheap computers were getting more powerful, and both were becoming more plentiful. Now data could be collected with a simple minded microcomputer, and communicated rather than carried to a larger system where the programming facilities and peripherals were more readily available for analysis, storage, printing of reports and plotting.

Today the personal computer with its own inexpensive peripherals and program development facilities has entered the scene. Data collection is easy and data reduction, analysis, reporting and plotting can often be accomplished with generic commercial packages eliminating the need for the scientist to program at all, something which most often he was not very good at anyway.

The most recent extension of instrument automation has come with the availability of practical laboratory robotics systems. These systems can be as easy to implement as the personal computer data system and extend automation beyond control, data collection and analysis for a single instrument or method to the automation of the entire analysis process including complex sample preparation. The robotic system is capable of automating not just a single analysis, but the total range of analyses required for each sample by a variety of instruments and methods simultaneously. In addition to this, several vendors are now offering Laboratory Information Management Systems (LIMS) which allow automated instruments to be networked to a host computer for sample scheduling, analysis and reporting.

Task Automation

There we have it, if data collection and analysis can not be done now, it is usually because someone doesn't want it to be done. Where then are the new horizons in laboratory automation? We return to the concept of task automation. Task automation involves determining what it is we should be doing, and using automation to accomplish it efficiently. This is a restatement of the now familiar efficiency and effectiveness concept.

It must first be recognized that the scientist works on two sides of the laboratory. There is the bench side of the lab where the instruments are located and experiments are performed and there is the desk side of the laboratory where the data is analyzed, the reports and publications are written and the multitude of desk and information management tasks are performed. It is essential that these two sides be brought together in order to automate the total task. Automation of the bench side is now familiar, but the scientist must perform the "keeping track of", the communicating, the information storage and retrieval and all the other activities generally associated with an office worker. Since the scientist is spending a large percentage of his time performing as an office worker, it would seem only logical that he should be provided with the same sort of tools and resources found in the modern office environment. This is particularly true if one considers that this office work is not the scientist's primary function and should be done as efficiently as possible in order to allow more time for technical activities.

The provision of these office automation tools to the scientist must be done in a way which integrates the office activities with the lab activities. Global planning must be done for the implementation of a comprehensive system which includes laboratory instruments, robotics, office automation, graphics, molecular, reaction and other modeling tools, information retrieval and all the other computer resources required by the modern scientist.

To implement this global system efficiently a consistent user interface or a universal workstation through which all automated activities can be performed is required. This allows consistency in the scientist's interactions with the system and allows him to efficiently move between workstations in the course of his work.

The manager and the secretary must be part of the global plan as well. Information transfer and communication are required not only with other scientists in the laboratory, but with the administration and support functions as well. The manager must bridge between the technical and the business functions. To accomplish this he must have access to the same technical and office tools used by the scientist, and in addition he requires access to business and financial information, project management tools, and decision support systems. In a large organization, this means access to both the scientific computer systems and to corporate business systems. Again, this must all be accomplished from a single workstation at the manager's desk in order to be effective.

Summary

Automation of today's laboratory should no longer be viewed simply as instrument automation. The modern scientist is an office worker as well as a technical worker and must be given the computer tools to allow the integration of the total laboratory task. The yields to the companies which recognize this will be significant improvements in both the efficiency and effectiveness of their research function.

Recommendations for Further Reading

The following is a list of recent articles and other references which, in addition to the many excellent papers in this volume, may help to provide a more detailed insight to some of the concepts discussed in this paper.

1. "Managing the Electronic Laboratory: Part I", R. E. Dessy, Ed., Analytical Chemistry, 56, 725A (1984).

2. "Managing the Electronic Laboratory: Part II", C. Snyder, R. E. Dessy, Ed., Analytical Chemistry, 56, 855A (1984).

3. "Laboratory Computer Networks", S. A. Borman, Analytical Chemistry, 56, 413A (1984).

4. "Office Automation, A User-Driven Method", D. Tapscott, Plenum Press, New York, 1982.

5. "Integrated Laboratory Automation", J. G. Liscouski, American Laboratory, 17(1), 63 (1985).

6. "Computers Gaining Firm Hold in Chemical Labs", P. S. Zurer, Chemical and Engineering News, 63(33), 21 (1985).

RECEIVED November 14, 1985

# 2

# Economic Considerations of Laboratory Information Management Systems

Joseph H. Golden

Laboratory Management Systems, Inc., New York, NY 10010

> Laboratory Information Management Systems (LIMS) have become a widely recognized tool for increasing the productivity and quality of service of the analytical laboratory. Laboratory managers are increasingly being faced with the problem of determining the benefits that LIMS could offer their organization, and of relating those benefits to economic measures which can justify a system's purchase or development. This paper presents an overview of the economics of LIMS. It presents a rationale for identifying sources of economic value to be derived from LIMS, and for estimating their worth. It presents the various factors that contribute to the actual cost of a system and finally, it suggests financial analysis techniques which can be employed to justify system acquisition.

The Laboratory Information Management System (LIMS) has achieved wide recognition as a powerful tool for increasing the productivity and quality of service of the analytical laboratory. Commercially available systems have been presented which range from inexpensive microcomputer based systems to half-million dollar or more superminicomputer based systems. In addition many firms have already developed or acquired custom systems tailored to their specific needs. Some large scale custom systems utilizing mainframe computers have actually been in the multimillion dollar range. Literature describing some representative examples of LIMS technology is cited at the end of this paper to provide an overview of this technology for the reader (1-7).

## LIMS Functions

In general, LIMS can perform a basic set of functions which greatly facilitate the operation of analytical laboratories: They can provide for work scheduling, for status checking and sample tracking, for automated entry and processing of analytical test data, for

---

This chapter is adapted from ACS Symposium Series No. 261, Computers in Flavor and Fragrance Research, edited by Craig B. Warren and John P. Waldradt.

automated report generation, for laboratory data quality assurance, and for data archiving. In addition, they can provide management level reporting of work backlog, turnaround time and laboratory productivity, and frequently also provide billing and other administrative information whose compilation would otherwise impose a considerable clerical burden on the laboratory staff. Table I presents numerous examples of the functions that may be encompassed by LIMS although it is unlikely that all of these functions would be incorporated in any single system.

As the literature cited demonstrates, the technology for implementing LIMS is now well established. One of the major problems now confronting those responsible for bringing LIMS technology into their environment is the need to economically justify the acquisition or development of a desired LIMS. This translates into a need for quantitative measures of the economic impact that LIMS will have on the organization as a whole, and includes both the cost of laboratory operations and their relationship to corporate revenue. This economic view is required to establish reasonable objectives and budgetary guidelines as well as to estimate the economic cost to value relationship of any specific LIMS proposal.

LIMS in different laboratory environments

First, let us consider some ways in which LIMS technology can contribute to the organization utilizing it. To better understand how mission impacts the benefits to be accrued through LIMS, we will consider analytical labs with different missions.

R & D Laboratories. The analytical lab supporting R&D is principally involved in the development of new products and processes, the improvement of existing ones, and occasionally, the analysis of competing products. Testing is more frequently done by professional analytical chemists rather than technicians. The work is often non-routine and method development may be included as part of the analytical tasks. In a corporate environment, this lab is an overhead function with significant labor costs. Data archiving may be required for a research data base, for trend analysis purposes, or for legal requirements dealing with patent applications or litigation or for regulatory agency compliance.

The R&D lab, then, would benefit from LIMS technology exhibiting high flexibility, the ability to structure large empirical data bases, and the ability to support scientific and statistical investigations.

QA/QC laboratories. The QA/QC lab is responsible for the testing of feedstocks and raw materials, process intermediates, and finished goods, and may, in addition, be responsible for the development of standards for materials, processes, and procedures. The QA/QC lab is usually characterized by the routine, repetitive nature of its workload. Testing is primarily to specification and, where lot acceptance or rejection is involved, is often on a grade category or pass/fail basis. Data may be archived for compliance with regulatory directives and for analyses of trends in material or process performance.

Table I. Functions Encompassed by LIMS Technology

I. ANALYTICAL SUPPORT FUNCTIONS
    Data entry and automated instrument interface
    Computational support
    Analytical result report generation
    Data archiving and retrieval
    Method and specification storage and retrieval

II. WORK AND RESOURCE MANAGEMENT
    Sample log-in
    Receipt and label generation
    Work assignment and scheduling
    Worklist preparation
    Sample tracking and status reporting
    Backlog reporting
    Report approval and release
    Reagent inventory and preparation control

III. LABORATORY QUALITY ASSURANCE SUPPORT
    Audit trail generation
    Multiple analysis, blind sample and round robin tracking and variance reporting
    Automatic tolerance verification and limit checking
    Instrument calibration scheduling and tracking

IV. MANAGEMENT SUPPORT
    Lab productivity analysis
    Turn around time and customer service analysis
    Cost per analysis computation
    Equipment utilization analysis

V. BUSINESS SUPPORT
    Labor time charge entry and reporting
    Customer account charging and/or billing
    Inventoried product data for order entry processing
    QC test data for feedstock purchasing and vendor qualification
    Data for corporate databases for regulatory agency compliance reporting (EPA, OSHA, FDA, TOSCA, etc.)

The QA/QC lab, then, would particularly benefit from LIMS technology which would mechanize the collection and analysis of data from routine tests, which would assure and document adherence to appropriate test methods and specifications, and which would include automatic limit checking and pass/fail determination.

Commercial Testing Labs. The Commercial Testing Lab is a service organization whose product is its tests. Its profitability is dependent on providing a high quality of service while minimizing its own cost per test.

The commercial testing lab may not need the data base of the R&D lab nor the speed of the Quality assurance lab, but because of the immediate link between its operations and its economic success, a system would be deemed beneficial if it reduced cost per test, enhanced responsiveness to customers, or speeded the collection of receivables through automatic invoicing.

## The common laboratory management problem

Regardless of the laboratory's mission, however, laboratory managers are confronted with a common set of problems: <u>Increases in data volume</u> from increased use of smart instruments and from increased testing and record retention requirements imposed by EPA, FDA, OSHA, and other regulatory agencies; <u>constantly rising operating and material costs;</u> and ever <u>tightening constraints on staff and material expenditures.</u> These are manifested by increasingly <u>burdensome paper work, inefficient utilization of resources,</u> and <u>exasperating searches for misplaced samples and data.</u>

## The Economics of LIMS

While these problems seem of immediate enough import to the lab manager, the economic impact of the laboratory's operations on the organization as a whole must be examined to assess the value to the organization of a capital investment to remedy these problems.

R & D. Returning to our examples, The R&D lab, contributes to the long term profitability of the firm (rather than the short term cash flow) by developing and perfecting products and processes. While controlling the costs of R&D as a whole is important, the speed at which a specific analytical test can be completed is less important than the speed and success at which a project as a whole can be completed. This relates to the effectiveness of the lab at its overall mission. The ability of a R&D lab to quickly and successfully develop products and/or processes and if necessary to protect them through patent actions, may ultimately impact the firm's market share and its profitability.

QA/QC. The QA/QC laboratory, on the other hand, is concerned with the quality of the firms products and therefore can influence product costs and revenues (the cash flow associated with those products). It determines acceptance or rejection of raw materials and feedstocks and/or assesses their market value for the purchasing department. It frequently initiates the processing of claims against vendors providing raw materials which are below grade specification but are nevertheless used. The lab also may be responsible for process monitoring to determine process parameters which minimize the production of scrap and off-grade product. The economic impact of LIMS technology in the QA/QC lab is primarily produced by its ability to speed the delivery of dependable information to those responsible for making immediate decisions regarding the purchase, production, and sale of product. Since the QA/QC lab budget may be relatively small compared with product production costs as a whole, the improvement in laboratory staff productivity which LIMS also offers, may be, in contrast to LIMS' importance an the R&D environment, actually of secondary importance to its impact on product costs and revenues.

Commercial Testing. The commercial test lab is by definition, in the business of testing. The satisfaction of its customers is paramount. Quality service and responsiveness holds old customers and attracts new ones. Getting dependable data fast is important, but here cost is critical. In a competitive market, the ability of LIMS to bring down the effective cost per test means more profit and/or the ability to keep or increase one's market share by lowering prices.

The Financial Perspective. Having looked at some general ways in which LIMS can enhance a lab's contribution to the corporate bottom line, we must next consider how this can be put into specific financial terms. In this process it will become apparent that the establishment of reasonable expectations for LIMS' economic benefits is to a great part dependent on understanding the mission of the lab rather than on the technical merits of the system (i.e., its ability to interface with various equipment or to perform specific functions desirable at the bench level). We will see, in fact, that the economic justification for acquiring LIMS technology must be based on an understanding of how the laboratory contributes to the corporate bottom line. This holds true whether one is trying to determine the scope of a LIMS that is appropriate, or to justify the acquisition of one being considered. In other words, we must consider not just what LIMS will do, but what it's worth.

Assessing the value of LIMS

To make that determination, we must examine the specific ways in which the analytical laboratory's product -- information -- contributes to the corporate bottom line as well as the ways LIMS can reduce the direct cost of laboratory operations. The intent here will not be to provide specific economic relationships, but to highlight the analysis approach.

LIMS value in the R & D lab. Returning now to the R&D lab, the economic value of LIMS is heavily skewed to the productivity area. This productivity improvement usually amounts to at least 10%-20% of total staff resources of a laboratory complex. The reader can verify this by assessing for his/her own labs the percent of staff activity spent manually recording data that could be captured automatically, transcribing and checking data, performing manual computations, searching for test data, looking up test methods and specifications, locating samples, consolidating data into issuable reports, and performing other clerical functions.

In addition, other economic benefits can be discovered although they may be harder to quantify. The speed at which a product can be perfected and brought to market may be the competitive edge through which the firm secures a lead position in its marketplace. Sometimes this means getting there first; sometimes it means fielding a product in response to a competitor — before the competitor has been able to capture the market.

Thus the rationale for LIMS technology in the R&D lab is not just an improvement in productivity or in the speed at which particular analyses can be completed, but also lies in the lab's ability to complete entire task projects in shorter time.

<u>LIMS value in the QA/QC lab.</u> In the QA/QC lab the primary area to investigate is the <u>time-value</u> of the information it produces. In the case of raw materials analyses, the ability to quickly accept or reject raw material lots may reduce costs associated with having to compensate for inferior materials or of holding materials or having to return them to their vendors. Similarly, the ability to quickly assess the quality and worth of raw materials in a competitive market may insure achievement of maximum value for the firm's purchasing dollar. In in-process testing, improving lab performance may contribute to process optimization or to reducing waste and rework. Here, the speed at which test data can be used to adjust a process can be directly related to the cost of the end product. This may be seen in the reduction of total production time and/or labor or in the reduction scrap or off-grade product, or in increasing the effective capacity (salable product per unit time) of the plant itself. Increasing effective plant capacity, for example, might reduce the need for finished goods inventories to accommodate peak demands. This would reduce the cost of inventory space, management, and insurance (it would also result in customers receiving fresher products where this is a factor). Furthermore, in applications where certificates of analysis are required prior to shipping or acceptance, time saved may translate into material holding and/or labor savings (e. g., for truck driver's or tanker crew's idle time awaiting authorization to depart).

To this must be added the 10% to 20% savings in direct lab productivity to be achieved through elimination of repetitive clerical work and through computerization of inefficient and error prone manual procedures.

<u>Commercial Testing.</u> Once again, in the commercial test lab environment, the ultimate test for LIMS is whether it makes the lab itself more profitable without regard for how its product, information is being used by others.

Cost Analysis

Having taken an overview of where benefits of LIMS may originate, and how they may differ depending on the mission of the lab using the LIMS, let us consider what these systems may actually cost.

First of all we must look at the total life cycle cost, not just the price tag on a purchased system or the development cost of an "in-house" system. This includes the costs of preparing a valid specification and/or requirements analysis (8), of site preparation if required (raised floors, air conditioning, etc.), of system purchase or development, of installation (including cabling), of integration into operations (including training and redundant activities during cutover), of continuing operation and maintenance, and perhaps even of insurance costs. If the system is developed in house there is the obvious cost of the development labor, but even if the system is purchased, staff will have to be committed to requirements analysis, liaison with the system developer/vendor, integration of the system into operations, and in-house system support and maintenance functions.

If the system is modest and stand-alone (perhaps based on a small minicomputer or super microcomputer), only the acquisition cost may be significant. If the system needed is a large and expensive one (such as would be based on a mainframe or super minicomputer); and if it requires specially prepared and/or air conditioned facilities, much cable installation and/or on-going service contracts, then all cost factors may require consideration.

Cost influencing factors. In estimating the real cost of a LIMS, however, the actual cost may be significantly less than the system's purchase price or apparent development cost. In estimating the true cost of a system one must consider many factors:

First of all, there is an immediate Investment Tax Credit which is currently 10%. This means that a system whose price is, for example, $100,000.00, would actually cost the firm only $90,000.00. Then using an accelerated depreciation rate, the system's cost can be written off over 3 or 5 years. If the firm in our example is in a 50% tax bracket, as is typical, the actual savings would be $10,000 on the ITC, plus 50% of the capital value, $95,000 (cost less one-half the ITC as per the current tax code) or a total saving of $57,500. This means that the $100,000 system would cost only $42,500 over a three year period. Now suppose that the firm borrows the initial cash outlay to buy the system and is paying a rate of 13% interest. The interest payments are also deductible at the firm's tax rate so the financing cost, offset by its tax deduction, raises the total system cost to roughly $70,000. The exact figures would depend on the details of the firm's fiscal policies such as whether payments are made monthly, quarterly, etc. whether purchases are financed or bought out of cash reserves, and other factors. The most salient point is that the true cash cost of a capital acquisition to the firm is usually between one-half and three quarters of its price tag.

The need to compare costs with benefits. Having looked at the real costs of a system one must then be prepared to answer that question, "is it worth it?"

This is where financial analysis comes into play. It is the process through which one decides whether the time to computerize the lab has come, and if so, how much or how little cost is justified by the anticipated benefits. It is also the process by which one can access whether or not a particular system being considered is worth its price.

Returning to our discussion of benefits, we note that they fell into two principal areas: the time-value of information and the improvement in laboratory productivity. As we saw, assessing the benefits from the time-value of information requires examination of the total process through which that information will be used. For example, using data such as the anticipated improvement in laboratory turn around time, the rate at which material is produced and its value, and the current the financial losses attributable to scrap or rework, one can estimate the savings in dollars to be accrued (each hour of faulty production avoided through faster lab turnaround time yields as a gain to the firm, the revenue that otherwise would have been lost).

Similarly, knowing the average labor, overhead, and General and Administrative (G&A) costs allows one to convert productivity improvements in the lab to dollar values of equivalent labor made free for more productive uses.

Thus, by understanding your lab's mission, and how it contributes to the organization's bottom line it is possible to develop quite reasonable and persuasive estimates for the benefits achievable through LIMS.

Having developed estimates of benefits as well as costs, these values must be compared. Since costs and benefits are not necessarily experienced in the same time frames, the same financial analysis methods used by the firm to make its other capital investment decisions must be used to justify the LIMS acquisition. The LIMS which is desirable from the technical and operational viewpoint can then be shown to be a truly beneficial capital investment and not just another corporate overhead expense.

Application of financial analysis methods

In general, one or more of three methods are used to justify major expenditures. The first, payback, is a measure of the time it will take for cumulative benefits to equal cumulative costs (time to break even). This, by itself, may not be sufficient to compare alternative investments and projects competing for the same limited resources so one of two other methods may be used. These methods, Net Present Value and Internal Rate of Return, consider the earning power of money in making comparisons. Because investments earn compound interest, a dollar to be gained in the future has less present value than one gained today. The NPV is computed by estimating the yearly cash flow (benefits less costs) generated by the acquisition over its life and then adjusting each years contribution to the total to its current value based on anticipated interest rates. An NPV of zero means break even, a greater NPV means that profit is returned. Similarly, IRR is an estimation of the equivalent compound interest earning rate of the investment over its life. Since financial managers are responsible for attaining the maximum return on their dollars, IRR is an important tool for comparing competing projects and investments. It is frequently used when available funds for major capital projects are tight.

Table II demonstrates how NPV would be calculated for a hypothetical LIMS, purchased as a package with negligible site preparation and with installation costs included in the purchase. It is to be acquired for a service laboratory primarily supporting R&D activities but with some minimal process monitoring responsibilities. The IRR for this project could be found by trial and error determination of the yearly discount rate which results in a zero NPV. A succinct discussion of these financial management analysis tools can be found in two works by Weston and Brigham. The first ([9]) presents theoretical and detailed analytical expositions; the second ([10]) is a more practical, applications oriented presentation.

Conclusion

All this discussion may seem a far cry from the problems of running a laboratory or of introducing LIMS technology to it. Indeed, if the lab is small, and its needs modest, all that may be needed to justify acquisition of LIMS technology is the general knowledge of LIMS' value in facilitating laboratory management and in increasing the speed, quality of service and productivity of the lab, and some realistic estimates of the system cost and payback.

On the other hand, if we are speaking of a large laboratory complex whose problems would necessitate a comprehensive computerization program costing from a quarter to well over three quarters of a million dollars, or perhaps to many millions of dollars, as some systems have cost, we may be dealing with a very different decision making process.

For such major system acquisitions, a more thorough analysis of the system's economics will probably be required. In such a case it is best if those responsible for exploring and/or securing the benefits of LIMS technology, understand the financial decision making process that goes on in the firm and can assess for themselves the value of anticipated benefits and the investment which such benefits would justify. Not only will this insure a judicious system acquisition, but it will develop, in advance, the specific data that the financial decision makers require and which proves the value of the system not just to the labs, but to corporate organization as a whole.

Having familiarity with the tools of financial analysis, those seeking the introduction of LIMS technology into their laboratory environments will be well prepared if such financial considerations are the ultimate arbiters of budget approval decisions. A

Table II. Net Present Value – Discounted Cash Flow Analysis

Assumptions:    Cost of LIMS is $250,000 both depreciated and amortized over 5 years.

Cost of money interest factor (I) is 13% and the firm is in a 50% tax bracket.

Lab has staff of 60 which, with overhead, costs a total of $3,600,000 per year.

LIMS will produce productivity savings of 5% the first year, 10% the second, and 15% thereafter.

LIMS will produce savings from expedited QC testing -- reduction of rework and scrap, improved claims processing on out of spec raw materials, and lower inventory maintenance and insurance -- of $25,000 the first year $50,000 the second, and $100,000 thereafter.

One senior staff member will be committed to the project full time during the first year. Operations and Maintenance (O&M) will cost $25,000 annually thereafter.

Formulae:    $PVCF(n)$ = Present Value of Cash Flow for year n
= $CF(n) * PVIF(n)$ where $CF(n)$ is the n-th year's Revenue less Costs and

$PVIF(n)$ = Present Value Interest Factor
= $(1+1/i)^n$ where n = the period (year) and i = the discount rate

$NPV = PVCF(1) + PVCF(2) + \ldots -$ Initial Cost

| CF Item | | Year 1 | Year 2 | Year 3 | Year 4 | Year 5 |
|---|---|---|---|---|---|---|
| Amortization of principal | (−) | 50,000 | 50,000 | 50,000 | 50,000 | 50,000 |
| Interest | (−) | 32,500 | 26,000 | 19,500 | 13,000 | 6,500 |
| Project labor | (−) | 60,000 | 25,000 | 25,000 | 25,000 | 25,000 |
| Tax Credits | (+) | 41,250 | 13,000 | 9,750 | 6,500 | 3,250 |
| Depreciation | (+) | 47,500 | 47,500 | 47,500 | 47,500 | 47,500 |
| Productivity | (+) | 180,000 | 360.000 | 540,000 | 540,000 | 540,000 |
| Revenue Gain | (+) | 25,000 | 50,000 | 100,000 | 100,000 | 100,000 |
| Cash Flow | | 151,250 | 369,500 | 602,750 | 606,000 | 609,250 |
| PVIF | | 0.8850 | 0.7831 | 0.6930 | 0.6133 | 0.5428 |
| PVCF | | 133,856 | 289,355 | 417,706 | 371,660 | 330,701 |

NPV = 1,543,278 − 250,000 = $1,293,278; PAYBACK in second year presentation of sound economic considerations added to the presentation of qualitative and technical benefits to be derived from LIMS technology can be decisive in the decision making process.

Literature Cited

1. Dessy, Raymond E. [ed], "Laboratory Information Management Systems: Part I"; Analytical Chemistry, Vol. 55, No. 1, JAN 1983
2. Dessy, Raymond E. [ed], "Laboratory Information Management Systems: Part II"; Analytical Chemistry, Vol. 55, No. 2 FEB 1983
3. Harder, Mark E., and Koski, Peter A., "A Microprocessor-based Scientific database Management System"; American Laboratory, SEPT 1983
4. Liscouski, Joseph G., "Distributed Laboratory Data collection and Management"; American Laboratory, SEPT 1983
5. Ouchi, Glenn I., "Lotus in the Lab"; PC World, Vol. 2, No. 2 FEB 1984
6. Reber, S., "Laboratory Information Management Systems"; American Laboratory, FEB 1983
7. Stinson, Stephen C., "Lederle's Toxicology Lab is Computerized Marvel"; C&EN, JAN 24, 1983
8. Golden, Joseph H., "Computerizing the Laboratory: The importance of System Specification"; American Laboratory, NOV 1980
9. Weston, J. Fred, and Brigham, Eugene F., Managerial Finance, Seventh Edition, The Dryden Press, Hinsdale, IL, 1981
10. Weston, J. Fred, and Brigham, Eugene F., Essentials of Managerial Finance, Third Edition, The Dryden Press, Hinsdale, IL, 1974

RECEIVED February 25, 1986

# Applications of Computer Data Base Management in Polymer and Coatings Research

Mark E. Koehler, A. F. Kah, and T. F. Niemann

Dwight P. Joyce Research Center, Glidden Coatings and Resins, SCM Corporation, Strongsville, OH 44136

>   The recently acquired ability of the scientist to have
>   easy access to uncomplicated computerized data base
>   management utilities without the necessity of, and the
>   constraints connected with, involving systems personnel
>   in these projects has provided the opportunity to store,
>   retrieve and analyze data in ways which would not have
>   been practical or even possible before.  The data base
>   applications range from very large, complex systems
>   which may have a life of many years, to fairly small,
>   short-term, personal data bases.  This paper attempts to
>   describe a range of applications of data base management
>   implemented in our laboratories.  These applications
>   include managerial applications such as project
>   monitoring and patent disclosure tracking; information
>   retrieval applications such as a research report index,
>   lab notebook tracking and file indexing; and technical
>   data retrieval and analysis applications involving
>   exposure test data and laboratory experimental data.

Computer Data Base Management Systems (DBMS) are not new.  DBMS has been used on large mainframe computer systems for traditional business applications for many years.  Most simply stated, a DBMS is a program which allows the user to perform various standard operations involved in the storage, maintenance and retrieval of data stored in organized files on a computer without the need to write a program. Common functions available in DBMS packages include STORE, SORT, MODIFY, DELETE and SEARCH. Usually these packages have the ability to generate standard procedures or sequences of commands for frequently used operations and for the generation of formatted reports, but their most powerful feature lies in the ability of the user to query the system to retrieve information based on the contents of one or more of the data fields in the file, or to cross index information between different fields.
    Recently, with the revolution in mini and microcomputers and

disk storage, DBMS packages have become readily accesible to the
scientist. In addition, the DBMS packages themselves have become
easier to use for the casual computer user. This recently acquired
availability of easy-to-use, inexpensive computerized data base
management utilities to the researcher is having a very beneficial
effect on the manner in which research is conducted. The technical
staff now has the ability to store, retrieve and analyze data in
ways which would not have been practical or even possible before,
and all without the necessity for, and the constraints connected
with, the competition for the scarce resources of Systems personnel.
The nature and scope of these technical applications of DBMS
covers a wide range. This paper will attempt to give an overview of
several applications in the areas of technical administration,
information retrieval, laboratory, and 'personal' data bases. The
intent is not to provide technical detail about any particular
application, but to provide a base of ideas which may benefit other
researchers in their information handling applications.

With the exception of the Project Activity Monioring System
(PAMS) which was developed on an IBM system under the CMS operating
system using the FOCUS$^1$ data base package, all of the applications
discussed were developed and operate on a Digital PDP11/44
minicomputer running the RSX11M+ operating system using DATATRIEVE$^2$
as the information retrieval package. While DATATRIEVE is not as
easy for the novice or casual user to use for small applications as
are some other packages now available, we have been quite successful
with it in our research center.

In practice, these technical details of implementation are of
the least importance to the user. A multitude of packages are now
available for a wide range of computer systems and the selection of
a DBMS for a particular application should be based mainly on the
criteria of ease of use and the features required for the particular
application. Consideration must also be made of the operating speed,
both of the DBMS package and of the computer and its storage system,
particularly in the case where large amounts of data must be stored
and searched.

## Administrative Applications

The Project Activity Monitoring System (PAMS) was developed to give
technical managers accurate and timely information about research
project time charges. Research projects are coded by technology,
market, principal researcher, technical manager responsible, funding
entity, etc. Each department's manpower budget forcast is stored
for comparison to actual time charges. Project time charges are
entered from the technical staff's weekly time sheets. Time may be
charged to valid technical projects or to "unrelated" categories
such as vacation, illness, training etc. Over twenty standard
reports are generated monthly. On-line inquiries on the status of a
particular employee or project can be made at any time. Special
reports and inquiries can be obtained on request. The data may be
searched by any of the coding fields and can be sorted and
summarized as desired by the manager. This system has proven most
valuable not as a financial tool for billing time charges but as a
tool for technical management to better manage their projects.

The Patent Activity Data Base was developed to provide better tracking and follow-up of this critical activity. The course of a patent disclosure from the time it is first disclosed until a patent is issued or it is abandoned can be a complicated process stretched over a time period of several years. Monitoring the status of the patent disclosures of a large research organization over this period of time is a formidable task. The Patent Disclosure Tracking Data Base was developed to centralize the stored information on patent disclosures in a searchable and retrievable form. This data base consists of two data files illustrated in Table I. The first is the

Table I. Patent Disclosure Tracking Data Base

| Patent Description | Activity |
|---|---|
| Docket Number | Docket Number |
| Title | Activity Code |
| Inventor | Country |
| Attorney | Date |
| Keywords | Comment |
| Serial Number | |
| Patent Number | |

Patent Description file and contains the relatively unchanging biographical information describing the disclosure. Included in this file are the docket number assigned to the disclosure at the time of its submission, the title, the author the technology involved in the invention, a series of keywords, the U.S. Patent Office serial number assigned when the application is filed, the attorney assigned to the case, and the patent number when it is assigned. The second file is the Activity File and it contains records of standard activities and their associated dates and comments. This information is cross indexed to the first file by the docket number. The standard report contains all information and activities associated with each selected disclosure sorted chronologically. Custom searches and reports are available on request. This system has made the monitoring of patent disclosures as they progress toward becomming issued patents a much more managable process. In addition, historical trends of issued patents and abandoned or rejected applications can easily be studied.

Information Retrieval Applications

The product of a research organization is knowledge. The distilled and summarized form of this knowledge is in the form of research reports. A research organization must to be able to search and access its research reports easily and efficiently. The Research

Report Index Data Base was developed for this purpose. It consists of biographical information for each report such as authors, dates, project numbers etc., and a carefully selected set of keywords describing the report. Valid keywords are maintained in a thesaurus data base to insure consistency in their selection and application. This data base may be searched on any logical combinations of biography and keyword information. Another information retrieval application is the Laboratory Notebook Tracking data base. This data base simply allows the user to inquire as to which employee a given laboratory notebook has been asssigned, or to inquire which notebooks have been asssigned to a given employee.

In addition to those information data bases developed within the research organization, a number of commercial data base systems are now available. Some examples of these commercial data bases and the types of information available are shown in Table II. Additional

Table II. Commercial Data Bases

| NAME | CONTENTS |
|---|---|
| DIALOG | Technical Literature<br>Business Information<br>Patents |
| National Library of Medicine | Toxicity Data |
| CAS On-Line | Chemical Compounds |
| TECHNOTEC | Technology Available for Licensing |
| SDC-ORBIT | World Patent Index |

commercial databases with information about a range of technical and business topics are also available.

Laboratory

A major application of data base in our laboratories is in the tracking of coatings test exposure data. The two types of exposure records currently implemented are for exterior hardboard siding coatings, and for coil coatings. While these two classes of substrates and the coatings used for them are quite different, the basic structure of the data and reporting requirements are not. There are three general types of data contained in three data files. These are illustrated for the exterior hardboard exposure system in Table III. The first file contains the information about the substrate. This includes an identification number for the panel, the date, the location and type of exposure to which the panel was subjected, the manufacturer of the substrate, the trade name, and the results of a series of standard test to evaluate the properties of the substrate. This information allows correlations to be made between the performance of the coating and the particular properties

Table III. Hardboard Exposure Test Data Base

| Board | Coating | Exposure |
|---|---|---|
| ID Number | ID Number | ID Number |
| Date | Date | Date |
| Manufacturer | Manufacturer | 12 Tests |
| Location | Trade Name | |
| Trade Name | Formula Number | |
| 4 Tests | Application Method | |
| | Film Thickness | |

of the substrate. The second file contains information about the coating or coatings which have been applied to the substrate. The coatings data contains the description, manufacturer, lot number, color and any other descriptive data for the coatings on each panel, and allows for multiple entries corresponding to multiple layers of primers and topcoats. In the case of coil coatings, various pre-treatment preparations may be included. This information facilitates the study of any possible interactions or relations between coats. The third file contains the actual exposure test data. The exposure data is the most voluminous as the panel may be evaluated many times over a period of several years before it is retired. This is true both of accelerated and exterior exposure testing. Typical exposure test records contain ratings for overall appearance, color and gloss retention, chalking, cracking, blistering etc. The coating and exposure records are cross indexed to the substrate file by the panel identification number.

Another application of data base in the laboratory is in the tracking of analytical sample analysis requests. A fairly large number of Laboratory Information Mangement Systems (LIMS) are now commercially available. These generally include the logging and tracking of analysis requests as one of their functions. In the case of simple applications, it is often not necessary to purchase a comprehensive LIMS package designed for a large analytical lab if a DBMS is available. In our laboratories, two groups have implemented analytical sample logging and tracking systems in this manner. Information contained in these data bases includes the name and group of the person requesting the analysis, the dates on which the sample was submitted and reported, the charge number and the number of hours charged to the analysis, and a sample priority which is assigned at the time of sample submission and is automatically upgraded as a function of time if the request is still pending. Worksheets of pending sample requests sorted by priority are generated for each group performing the analyses. Reports of time charges and sample requests are generated to aid managers in manpower and resource management and planning.

## Personal

A number of individuals have developed small data bases for their personal job use. An example of a data base of this type in the computer group is one containing the records of which employees within the research center have attended various computer training classes. This file is used to generate notices to managers when a series of classes is being planned so that they can review the training records for their groups and decide who needs to attend which class. This data base is then updated at the completion of each course.

Another personal data base in this department contains the acquisition dates and maintenance records of all computer terminals and printers in the research center. Since this equipment moves around the building, these records have made it much easier to decide if a particular unit is in warranty or is having a recurring service problem.

## Conclusions

The benefits realized from the use of data base in research lie mainly in allowing the scientist to do the job of research better. Manpower reductions are usually not realized since time savings are generally reinvested. What generally is realized are gains in technical productivity and in the quality of work. In the case of the exposure test data bases, the development groups now are able to get the full value of their exposure data due to its accessibility. Duplication of effort which was required simply because data could not be located has been virtually eliminated. There has also been enhanced credibility with customers because data can be produced quickly and completely. In a chemical coatings business where other factors are often nearly equal, this translates to increased sales. In addition, products are being brought to market faster with increased confidence in warranty specifications which again translates to increased sales.

In summary, database capability may be the single most valuable computer tool which can be provided to research today.

## Notes

1. FOCUS is a trademark and product of Information Builders.

2. DATATRIEVE is a trademark and product of Digital Equipment Corporation.

RECEIVED November 14, 1985

# 4
# Advances in Scientific Software Packages

**Channing H. Russell**

**BBN Software Products Corporation, Cambridge, MA 02238**

> Early scientific software packages focused on compilers, individual applications, and specific aspects of computer support such as statistics. More recently, software packages provide a broad, integrated, easy to use, and extensible set of capabilities to support research data management. RS/1 (TM) is described as an example of modern scientific software.

While most publicity about the software industry focuses on applications in office and home settings, there is also significant development taking place in packaged software specifically designed for research data management. Use of commercially available scientific data management software, either as a stand alone tool or as the base for building applications, is increasingly contributing to the productivity of research organizations.

It is indicative of the growing importance of the scientific software market that a major corporation, Bolt Beranek and Newman Inc. (BBN), has recently formed a new business activity called BBN Software Products, focused entirely on developing software for science. BBN Software Products' leading software product, RS/1 (TM) (BBN, 1979), incorporates the concepts and technologies developed through BBN's history of support of scientific applications. RS/1 is an integrated information-handling environment supporting data management, analysis, graphics, statistics, modeling, and reporting. The package is used by a broad spectrum of scientists engaged in a wide variety of research and development activities.

Based on their corporate experience in supporting their own research and development projects as well as a number of government-sponsored activities, the development of scientific software at BBN grew out of a long history of support of a variety of scientific applications. For example, in the 1960's, BBN developed a hospital information system for Massachusetts General Hospital which used an early minicomputer, the PDP-1, to support clinical research activities. Under the sponsorship of the National Institutes of Health, they developed a system called PROPHET (1) which provides a national timesharing service to scientists studying the effects of chemical substances on biological activity. PROPHET operates as a kind of automated laboratory notebook, oriented around data structures such as tables and graphs which are familiar to scientists (Castleman et al., 1974). A similar orientation towards data representations familiar to scientists is used in the CLINFO system, which provides specialized support for the study of time-oriented clinical data in the general clinical research center setting (Gottlieb et al., 1979).

RS/1 represents a distinct contrast to earlier scientific software. In general, early software used by scientists took three primary forms. (1) Languages such as FORTRAN, BASIC, and PASCAL were helpful in supporting highly computational applications, but were not designed to support research data management. (2) Specific scientific programs were also developed which were limited to supporting a single narrow applications area and were usually operable only using particular computer system equipment. (3) Specialized software packages began to appear in specific areas, such as SAS (SAS, 1982) and BMDP (Dixon, 1975) in statistics and MLAB (Knott, 1979) in curve fitting.

Some of the earlier software, such as SAS or APL, has since been extended to provide broader support, often with limitations that reflect their history. In addition, integrated business packages, such as Lotus 1-2-3 in the microcomputer world and some of the data management packages, have also been used for scientific work.

BBN took a more systematic approach than had been used in the development of earlier software packages for the scientific market; the company's goal from the start of the development process was to design an integrated general purpose package to provide broad-based interactive support for research data management. The result was RS/1.

## Characteristics of Modern Scientific Software

In order to understand the characteristics of state-of-the-art scientific software, a description of the facilities of the RS/1 system is presented in this section as an example of the kind of software now available to support scientific data management needs. All facilities are based in a single integrated system. No extra steps are needed, for example, to use tabular data as the basis for statistics or to graph the results of modeling.

## Data Management

Data management in RS/1 is based on two-dimensional tables. Each cell of a table can contain data representing fixed or floating point numbers, dates, times, or free text. Cells in a particular column are not all constrained to the same type; it is possible, for example, to include a note about some missing data in a column of numerical results. A user can work with many hundreds of tables. Tables are based on disk files, accessed through a kind of paging scheme, so there is no limit on table size. Some users work with tables containing hundreds of columns and tens of thousands of rows.

## Data Analysis

Data analysis makes direct use of tables. A new column can be created as a transformation of existing data. No additional steps are needed to create the column; a single command defines the transformation and transfers the data (Figure 1). The results can be seen on the screen immediately, again without additional commands or setup steps.

## Graphics

RS/1 supports the major kinds of analytical graphics needed in a scientific setting, including scatterplots, fitted curves, bargraphs, histograms, piecharts, three-dimensional displays, (Figure 2), and contour plots. Graphs are stored as permanent data objects that can be edited and redisplayed. This makes it possible to transform a graph made for analysis into a graph suitable for publication. An advanced terminal independent graphics support capability permits the output of pictures on more than a hundred different graphics devices.

## Statistics

Statistics available in the system include a large set of commonly used analysis techniques, as well as advanced nonlinear curve fitting techniques. Statistical results can be displayed numerically or graphically.

## Modeling

Modeling offers a spreadsheet-like capability, which permits the integration of several different spreadsheets and the inclusion of general data manipulation commands in cells.

## Text and Graphics

An optional extension to the system makes it simple to produce documents containing mixed text and graphics illustrations, which can be output on a laser printer.

## Easy to Use

Ease of use is a major theme in modern commercial software. The best packages offer several styles of interaction suitable for advanced or beginning users. In RS/1, both a command based and a menu based style of interaction are available. Using an English-like command, a bargraph can, for example, be constructed from a table in a single step (Figure 3).

Curve fitting can be accomplished in much the same straightforward way (Figure 4). Menu oriented interactions are also supported to make complex editing easy to understand, and to provide a way for beginning users to start using the system. Thorough documentation and optional direct and videotape training are available.

Drill Test Data

| 0 Bit Code | 1 Hardness | 2 Feet Drilled | 3 Final Height | 4 Lower Bound | 5 Upper Bound |
|---|---|---|---|---|---|
| 1 N22 | 0.80 | 218 | 0.50 | 0.450 | 0.550 |
| 2 N32 | 0.84 | 309 | 0.48 | 0.432 | 0.528 |
| 3 N44 | 0.80 | 662 | 0.41 | 0.369 | 0.451 |
| 4 N81 | 0.80 | 512 | 0.46 | 0.414 | 0.506 |
| 5 N82 | 0.84 | 114 | 0.59 | 0.531 | 0.649 |
| 6 S14 | 0.82 | 319 | 0.49 | 0.441 | 0.539 |
| 7 S17 | 0.80 | 404 | 0.47 | 0.423 | 0.517 |
| 8 S19 | 0.83 | 844 | 0.33 | 0.297 | 0.363 |
| 9 S22 | 0.83 | 56 | 0.74 | 0.666 | 0.814 |

Collected 4/18/82
Measurements of height +/- 10%

# COLUMN 5 OF DRILL = COLUMN 3 * 1.1 <RET>

Figure 1. A single command allows the user to create a new column in an RS/1 table as a transformation of existing data. (Reproduced with permission from University of South Carolina Press: Columbia, S.C., 1986; Channing Russell In Research Data Management in the Ecological Sciences; Michener, William, Ed.; pp 373-381.)

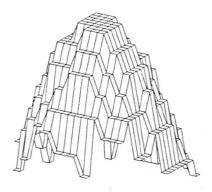

Figure 2. Graphs, like the 3-dimensional display shown here, can be stored as permanent data objects that can then be edited and redisplayed. (Reproduced with permission from University of South Carolina Press: Columbia, S.C., 1986; Channing Russell In Research Data Management in the Ecological Sciences; Michener, William, Ed.; pp 373-381.)

4. RUSSELL  *Advances in Scientific Software Packages*  27

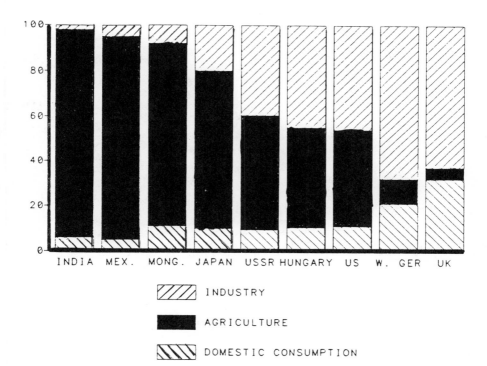

Figure 3. A bar graph can be constructed from a table in a single step. (Reproduced with permission from University of South Carolina Press: Columbia, S.C., 1986; Channing Russell in Research Data Management in the Ecological Sciences; Michener, William, Ed.; pp 373-381.)

## BIT WEAR ANALYSIS

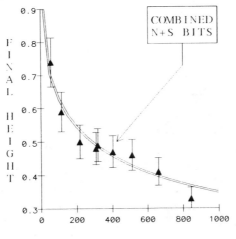

▲ Final Height

═══ $1/(1+\mathrm{SQRT}(3.497648e-03*X))$

Drill Test Data

| Bit Code | Hardness | Feet Drilled | Final Height |
|---|---|---|---|
| 1 N22 | 0.80 | 218 | 0.50 |
| 2 N32 | 0.84 | 309 | 0.48 |
| 3 N44 | 0.80 | 662 | 0.41 |
| 4 N81 | 0.80 | 512 | 0.46 |
| 5 N82 | 0.84 | 114 | 0.59 |
| 6 S14 | 0.82 | 319 | 0.49 |
| 7 S17 | 0.80 | 404 | 0.47 |
| 8 S19 | 0.83 | 844 | 0.33 |
| 9 S22 | 0.83 | 56 | 0.74 |

Collected 4/18/82
Measurements of height +/- 10%

Figure 4. Users can fit curves to the data points in an RS/1 table by entering a simple English command. (Reproduced with permission from University of South Carolina Press: Columbia, S.C., 1986; Channing Russell In Research Data Management in the Ecological Sciences; Michener, William, Ed.; pp 373-381.)

## Extensibility

A key attribute of software intended for use in science is extensibility. Research data management, almost by definition, involves special purpose information handling beyond standard data management techniques. These specialized needs are often confined to one particular phase of analysis; ideally, scientific software should support the smooth integration of special-purpose programming into an application which also makes maximal use of what is already available in the system. Too often in the past, entire new systems had to be constructed in a language like FORTRAN because the existing higher-level data management systems did not support the needed kinds of customization.

There are several kinds of extensibility built into RS/1. A full structured programming language called RPL is a part of the system. This language, stylistically similar to PL/I, also allows direct access to data objects such as tables and graphs, and allows the intermixture of high-level data management commands with traditional programming constructs.

The RPL language is designed for easy-to-write, compact programs. Like APL, it supports a run-time environment in which variables can represent different data types at different times. There is no need for the kind of data declarations which make programming awkward in traditional languages.

## Flexibility

In the scientific world, maximum flexibility is needed in interfacing to a variety of different programs, in accessing various databases, and in outputting information to different kinds of graphics devices. This goes far beyond the kinds of flexibility needed in business-oriented packages. RS/1 supports an ability to call programs written in other languages, and has interfaces to a growing set of commercial data base systems. Especially powerful applications can be constructed using information from a large database as the basis for analysis.

Output can be directed to a variety of printers, plotters, and display terminals, and it is even possible for sophisticated users to add support for new kinds of devices, through a facility called the terminal data table, without a need for system programming.

## Applications of Scientific Software

Scientific software packages are beginning to have a large impact on productivity in research organizations. In a major pharmaceutical research and development organization at Merck, Sharp and Dohme, RS/1 is used at several different laboratories in the U.S., Canada, the U.K. and continental Europe to share data. A large VAX installation supports users in different sites through a computer network. Collaborative projects between different laboratories, using data shared through the central system, have become common.

The semiconductor industry makes heavy use of manufacturing data to provide ongoing quality control information. RS/1 is used as the central software tool to tie together a number of different data bases and software systems to make the data available for graphing and analysis.

More than a thousand scientists and engineers use RS/1 at DuPont's Experimental Station in Wilmington, Delaware. It was particularly attractive to DuPont that the software was available on microcomputers as well as superminicomputers; this enabled them to adopt RS/1 as a standard for remote research labs and engineering stations in production plants throughout the Eastern seaboard.

In the new field of genetic engineering, scientific data management software is used to manage the long alphabetic codes that represent genetic sequences, as well as more traditional numeric and text applications. At Genentech, scientists use the software for these tasks as well as for laboratory data analysis.

Future Developments

Advances in computer science continue to serve as the basis for new extensions to software products. In particular, artificial intelligence techniques have begun to mature to the point at which they can play a role in scientific software. In the future, scientific software will incorporate expert systems technology in order to provide a new level of assistance to scientists in applying statistical and graphical techniques to data analysis. Two major areas are likely to be the focus of expert systems in the scientific software area: assisting users without extensive statistical training in starting to use statistics, and helping design multifactor experiments.

Acknowledgments

The PROPHET system is sponsored by the Biotechnology Resources Program, Division of Research Resources, National Institutes of Health under contract #N01-RR-8-2118.

Literature Cited

Castleman, P. A., Russell, C. H., Webb, F. N., Hollister, C.A., Siegel, J.R., Zdonic, S.R., Fram, D.M., "The Implementation of the PROPHET System", AFIPS Conference Proceedings, Vol. 34, pp. 457-468, 1974.

Dixon, W.J., Ed. BMDP Biomedical Computer Programs, 3rd Ed., Los Angeles, University of California Press, 1975.

Gottlieb, A.G., Fram, D.M., Whitehead, S.F., Rubin, G.M., Russell, C.H. , Castleman, P.A., Webb, F.N., "CLINFO: A Friendly Computer System for Clinical Research", XII International Conference on Medical and Biomedical Engineering, Jerusalem, Israel, 1979.

Knott, Gary D., "MLAB - A Mathematical Modeling Tool", Computer Progress in Biomedicine, Vol. 10, No. 3, pp. 271:280, December 1979.

SAS Institute Inc., SAS User's Guide: Basics, 1982 Edition, Cary, NC, 1982.

RECEIVED May 5, 1986

# Computer-Assisted Polymer Design

Rudolph Potenzone, Jr., and David C. Doherty

Chemical Research Division, American Cyanamid, Stamford, CT 06904

>The use of computers as an additional tool in studying polymer shapes is discussed. A general discussion of molecular modeling is presented with specific application to polymers. An example relating hydrodynamic properties of hyaluronic acid to computer generated models is included.

Computer aided design (CAD) is used in many diverse industries and technologies. The major benefits of CAD systems are: 1) gain in productivity; 2) the ability to evaluate more designs; and 3) the elimination of some prototypes. It is not surprising that in chemistry, computers continue to move into new areas. Academic programs have pushed computers into every aspect of their research including laboratory automation and data collection, data handling, statistical analysis and molecular modeling including quantum mechanics and molecular mechanics.

Chemists can now talk about using computer techniques to design new molecules, particularly in the pharmaceutical application of drug design. With bigger and faster computer systems, larger chemical systems can be examined. The result is that molecular modeling is becoming more useful to polymer chemists.

Molecular modeling tools have been used to study polymeric systems for over twenty years ([1],[2]). The availability of computer hardware and software has often limited the use and development of these methods.

Computer advances now make these tools as available to the polymer chemist as they are to drug designers. While it is conceptually possible to design a polymer de novo using computer modeling techniques, this remains an optimistic goal.

We shall use the Chemical Modeling Laboratory (CHEMLAB) (3) to illustrate the utility of computer modeling since this is the software that we have available in our laboratory. There are several other software packages that perform similar functions that could also be used (4).

CHEMLAB can provide information such as energetic feasibility, hydrogen bonding potential, etc. These can be used to explain observed behavior or to predict the properties of proposed compounds. 'Hypothesis testing' is the greatest utility of molecular modeling.

## MOLECULAR MODELING TOOLS

A wide variety of computational chemistry software and algorithms have been applied to polymers over the years. These can be classified in the following types:

Model Building. CHEMLAB offers a variety of ways to build a model and insert it into its molecular workspace. This includes structure calculation from standard bond lengths and bond angles as well as a true three-dimensional molecular editor. Molecules can be joined in three-space or built from three dimensional fragments.

Geometric Examination. The polymer chemist needs to examine the various characteristics of the molecule in the molecular workspace. Bond lengths, bond angles and torsional angles can be measured for the current structure and compared to accepted values. In addition, other geometric properties can be computed like overall dimension, moments of inertia, molecular volume and surface area.

Geometric Optimization. The structure of the molecule as built by CHEMLAB (or a input from other methods) can be optimized through either a full force field molecular mechanics calculation (MMII) or with the semi-empirical molecular orbital methods MINDO-3 and MNDO.

Molecular Energetics. Molecular energies can be computed in a variety of ways including empirical fixed valence potentials, full force field potentials, and semi-empirical molecular orbital techniques (CNDO-2, INDO, MINDO-3, MNDO, PCILO).

Conformational Analysis. Conformational flexibility is of particular importance for polymer models since it is often here that many of the bulk properties originate. The traditional analysis including a sequential SCAN, gradient search minimization and its ROTAMER algorithmic search can be useful in polymer modeling; these are generally inadequate for polymer modeling due to the large number of freely rotating bonds in even small polymer fragments. PSCAN, a customized CHEMLAB polymer modeling option, can be useful for homopolymer modeling, since it allows you to equate bonds in each monomer, and then SCAN only over the bonds in the initial monomer.

Molecular Probe Analysis. In an effort to understand how a molecule is 'seen' by either another molecule or by a surface, molecular probes can be moved around a chemical to map out its surface. These probes include anions and cations (point charges) and 'hard' spheres or can be constructed as a combination of these. The empirical potential energy is computed at a variety of points around the test molecule and an energy surface is thus generated. This can be examined graphically and compared as changes are made to the molecule.

Molecular Shape Analysis. Once a set of 'shapes' or conformations are generated for a chemical or series of analogs, the usual question is which are 'similar'. Similarity in three dimensions of collections of atoms is very difficult and often subjective. Molecular shape analysis is an attempt to provide a 'similarity index' for molecular structures. The basic approach is to compute the maximum overlap volume of the two molecules by superimposing one onto the other. This is done for all pairs of molecules being considered and this measure, in cubic angstroms, can be used as a parameter for mathematical procedures such as correlation analysis.

Intermolecular Energetics. Molecular calculations have long been criticized as unrealistic in considering only isolated molecules. Many examples now appear in the literature with molecular mechanics computations on two or more molecules (5). These types of calculations are very time consuming and difficult to analyze. Still, some excellent progress has been made on two or more interacting molecules using techniques such as those available in CHEMLAB.

Molecular Graphics. The first contact many chemists have had with computer modeling is usually with computer graphic representations of chemicals. These pictures are providing much better insight as to the spatial relationships with and between molecules. CHEMLAB offers a range of color graphic representations of a molecule in the molecular workspace. Some of the other modeling systems offer real time color graphic displays that can give dramatic presentation of the polymer models.

## POLYMER MODEL BUILDING

We have designed PBUILD, a new CHEMLAB module, for easy construction of random copolymers. A library of 'monomers' has been developed from which the chemists can select a particular sequence to generate a polymeric model. PBUILD takes care of all the atom numbering, three dimensional coordinates, and knows about stereochemistry (tacticity) as well as positional isomerism (head to tail versus head to head attachment). The result is a model of the selected polymer (or more likely a polymer fragment) in an all trans conformation, inserted into the CHEMLAB molecular workspace in literally a few minutes.

This model can then be examined graphically in PBUILD, added to from other monomers or other polymer fragments, or subjected to any of the other CHEMLAB functions.

We have added a companion option to PBUILD, PRANDOM which eases considerably the problem of finding 'good' conformations of a polymer segment. PRANDOM automatically selects all of the polymer backbone and/or side chain bonds and will randomly select rotations for each bond. In a few minutes, one can not only build a polymer fragment, but also set up a Monte-Carlo search of its conformational space. However, even this cannot solve the problems for large models (pentamer or larger), again due to the number of bonds to be rotated.

In order to be truly useful, it must be possible to compute polymer chain information, namely average mean squared end-to-end distances. These can be directly compared to polymer solution data if available and also provide information on bulk properties.

A method for estimating the chain statistics of a random monomer is available if the monomer conformation energetics are known. It allows for varying monomer arrangements (given a monomer composition) and generates energetically valid statistical chains, estimating the mean squared end-to-end distance. This is performed randomly after the user only specifies the names of each of the monomers (from the monomer library), their percentage in the final polymer, the length of the generated polymers, and the number of these polymers over which to average. For each polymer generated, the user can see the exact sequence selected, the rotation angles, the total energy and the end-to-end distance. The latter is saved for the averaging which is reported at the end of the procedure.

We used this procedure in estimating the chain properties of hyaluronic acid (6). Hyaluronic acid is a regular polysaccharide with a disaccharide repeat unit. The two units are glucuronic acid (Figure 1) and N-acetyl glucosamine (Figure 2).

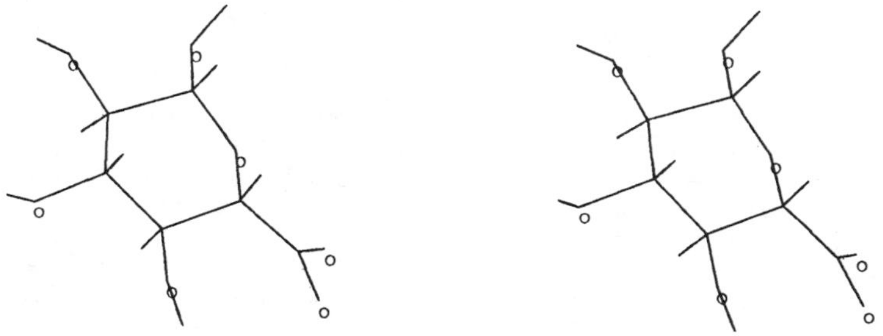

Figure 1a. Stereo image of Glucuronic Acid saccharide.

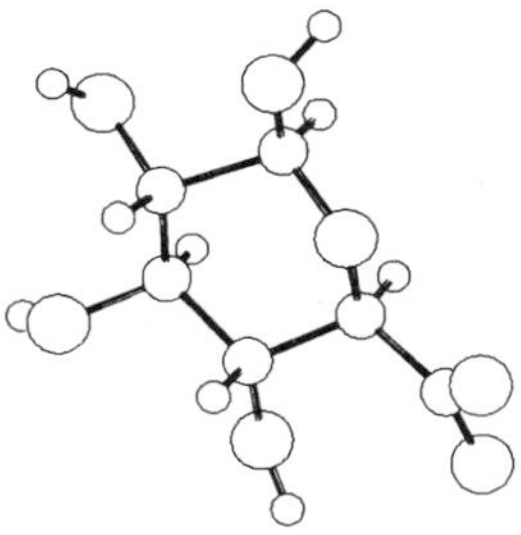

Figure 1b. Ball and Stick image of Glucuronic Acid saccharide.

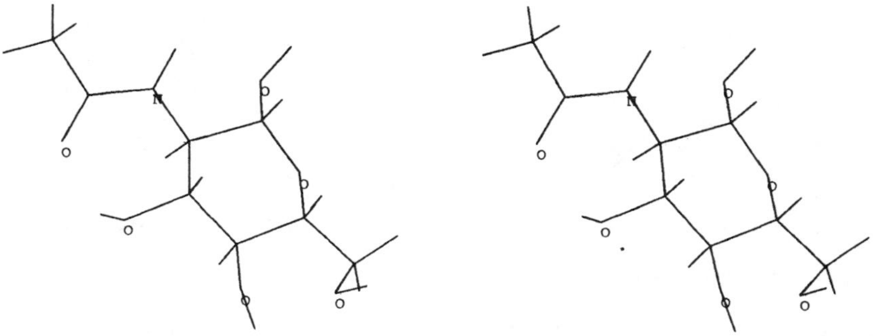

Figure 2a. Stereo image of N-Acetyl Glucosamine saccharide.

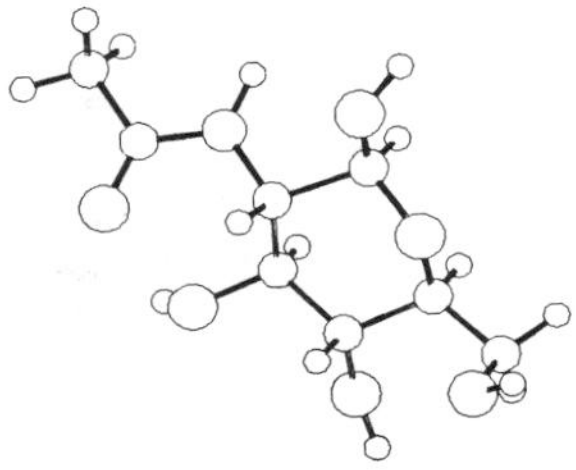

Figure 2b. Ball and Stick image of N-Acetyl Glucosamine saccharide.

Once the energetics of these monomers were studied, larger fragments were generated and thermodynamic probabilities of allowed conformational states were generated. These probabilities were used as weighting factors in generating **statistical chains** of hyaluronic acid. Hundreds of thousands of such chains were generated and average chain statistics were calculated. Very good qualitative agreement with experiment was found along with surprising quantitative agreement (6). Figure 3 shows a typical trimer (i.e. a hexasaccharide).

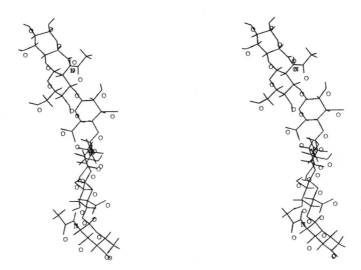

Figure 3. Stereo image of a Hyaluronic Acid trimer.

## MODELING AT THE BENCH

As with many new techniques available to the research chemist, it is hard to decide who should 'run' the theoretical data collection. Many drug companies have set up computer assisted drug design groups who provide, almost on a service basis. Molecular modeling, however, should be an easily accessible tool to those who need the information that it can provide.

There is no substitute for the chemist's intuition. The bench chemist should personally perform the primary analysis of the theoretical data and possibly the data generation itself, as long as the chemist understands that this method is not a miracle solution (and will not replace him or her). Ownership of the problem will often mean that the bench scientist is eating, sleeping and drinking the problem. This is the person who has the most insight into the problem and this is the person you want to look at the molecular graphics and the probe surfaces.

We suggest that approaches like those in CHEMLAB can be most effective as a thinking aid for the chemist. As Stephen Wolfram said in the June 1984 Scientific American (an excellent issue on computers) (7), "Computation is emerging as a major new approach to science, supplementing the long standing methodologies of theory and experiment ... Computers extend the realm of experimental science."

This extension in the laboratory can be seen as the fantastic 'hypothesis testing' application of molecular modeling. It is rare to find a chemical problem where there are not at least a few theories of the molecular mechanism involved. How many times has each of us heard 'steric affect' or 'hydrogen bonding' invoked as the explanation of a variety of experimental observations made at the bench level? How useful would it be to be able to actually build accurate, quantitative models to investigate such ideas?

## CURRENT USES OF MODELING

Molecular modeling potentially can be used in all design work of new chemicals to serve as monomers as well as in novel copolymers. Computer models provide the synthetic chemist with a realistic three-dimensional view of their compounds and allows for advanced 'hypothesis testing'. Rapid evaluation of steric effects and potential hydrogen bonding can be invaluable at the beginning of the design of a new chemical. In addition, chemical modifications can be done in minutes and changes in geometry and conformation easily examined. New methods to compare 'similarity' can provide useful insight in understanding either the synthesis or the reaction of a molecule.

## TRUE DESIGN OF MOLECULES THROUGH MODELING?

This remains a goal for the future. However, many groups are working on computer systems that will provide the tools needed to perform true design of new chemicals FIRST at the computer terminal.

In order for us to effectively develop and use these new tools, we must make the transition from an empirical, retrospective use of modeling to a planned design approach. The question to be addressed should not be: Why didn't this experiment work? Rather, we need a prospective outlook: Can this work? These new theoretical tools should be bringing new information to the chemist to be used in conjunction with experimental data already available. The success of computer aided design of chemicals will arrive when a chemist can sit at the terminal as the first step in the development process.

Literature Cited

1. Flory, P.J. "Statistical Mechanics of Chain Molecules"; Wiley Interscience: New York, 1969.

2. Hopfinger, A.J."Conformational Properties of Macromolecules"; Academic Press: New York, 1973.

3. **CHEMLAB-II**. The Chemical Modeling Laboratory. For information contact Molecular Design, LTD., 1122 B Street, Hayward, CA 94541.

4. Other software packages that are available for molecular modeling include: CHEMGRAF (Chemical Design Ltd., Oxford, England); SYBIL (Tripos, St. Louis, Missouri).

5. Carotti, A.; Hansch, C.; Mueller, M.M.; Blaney, J., J. Med. Chem. 1984, 27, 1401-1405.

6. Potenzone, R.; Hopfinger, A.J. Polymer Journal 1978, 10, 181-199.

7. Wolfram S. Scientific American Sept 1984, 251, 188-203.

RECEIVED December 13, 1985

# 6

# Silicone Acrylate Copolymers: Designed Experiment Success

**T. R. Williams and M. D. Nave**

**3M, St. Paul, MN 55144**

> Various statistically designed experiments and data analysis were used to investigate the properties of copolymers based on a new silicone acrylate oligomer (SiUMA). Both factorial and mixture designs were used, depending on the variables being investigated. Computers were used throughout the analysis of the data, and their use was essential to the timely investigation of many compositional variables. This paper presents the results and examples of output from computer software applied to this problem. It is shown that use of the principles of designed experiments and of the power of computers led to the successful identification of a material which satisfied many physical property targets simultaneously.

Copolymers of silicone acrylates (SiAc) and various hydrocarbon (non-silicone) acrylate monomers have been in use for some years (1) as materials for various diverse applications, including biomedical devices (2), release coatings (3), etching resists (4), and adhesives (5). A new SiAc oligomer has been developed at 3M with certain advantages over competitive materials. We desired to develop an understanding of the behavior of this new material by itself and in copolymers, but it was not known how the new oligomer would best be combined with hydrocarbon monomers to give interesting and useful combinations of properties. This problem is amenable to designed experimental techniques and response surface methodology(11-13), and this paper describes the successful application of these techniques to the problem's solution. The ready availability of both mainframe and personal computer programs, to handle the formidable mathematical manipulations involved in analysis and display of experimental results, was a key element in the successful outcome of this project.

This paper does not aim at educating the reader in how to use the concepts of statistically designed experiments, but rather at showing how computers can make the task relatively rapid and convenient. It is assumed that the reader has at least a passing knowledge

0097-6156/86/0313-0039$06.00/0
© 1986 American Chemical Society

of the concepts and nomenclature of elementary statistics and designed experiments.

## Experimental

The details of oligomer and polymer sample preparation will be published separately([6](#)). Briefly, the synthetic scheme is shown in Figure 1. Tetramethyldisiloxane (TMDS, Petrarch Systems, Inc.), octamethylcyclotetrasiloxane ($D_4$, Dow Corning), and isocyanatoethyl methacrylate (IEM, Dow Chemical) were used as received. The "dihydride" was prepared with controlled molecular weights by a known method([7](#)). The hydrosilylation of protected allyl alcohol was performed in accordance with well established practices([8](#)). The final product oligomer is a silicone urethanemethacrylate. In this paper the terminlogy "SiUMA-N" is used to denote such an oligomer, in which there are a number average of N silicon atoms per chain, where N = X+1 in Figure 1.

The cured polymer samples used for physical property testing were prepared by photocuring 12 mil thick sheets of degassed and photosensitized monomer mixtures, using a mold composed of glass plates lined with polyester film and separated by a double thickness of vinyl electrical tape. A GE sunlamp was used for illumination, and Darocure 1173 (E. Merck) was used as the photoinitiator. Hydrocarbon monomers were used as received from the manufacturers. All the vinyl group-containing compounds were stored at -5°C until use.

All numerical analysis reported in this paper was done using computer software of the type widely available throughout the country. MINITAB, licensed from Pennsylvania State University, was convenient and effective for entering data in matrix form, performing Yates and various regression analyses, and some graphics. However, the contour plots were done using a separate proprietary software package developed to aid in the creation, execution, and analysis of "designed" experiments. The Figures in this paper were generated with a mainframe-based graphics program using a personal computer as a smart terminal. Finally, this manuscript was composed and printed with a word-processing program on the personal computer.

## Results and Discussion

Table I is a list of physical properties of materials which were of special concern, along with target values felt to indicate useful levels in a particular application. From the beginning it was predicted that one of the biggest problems would be to balance Properties A and E, usually considered mutually exclusive. It was also assumed that Properties B and E were highly correlated. Statistically designed experiments and data analysis were chosen to determine most efficiently the formulations which would give the best combination of all the target properties.

The first design was intended as a "range-finding" experiment which would broadly identify promising ranges of the variables, which could in turn receive more detailed investigation later. This design was a $2^{4-1}$ fractional factorial design with variable D assigned to the ABC three-factor interaction[11]. This choice of design allowed the combination of three different variable types:

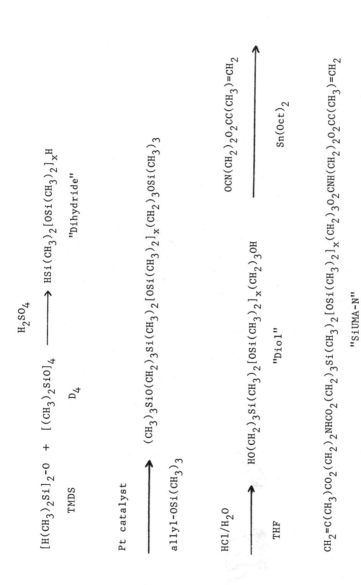

Figure 1. Synthesis of SiUMA oligomers.

SiUMA chain length, component ratios, and a component absolute percentage.

Table I.     Initial Property Targets

| PROPERTY | TARGET |
|---|---|
| Property A | >15 |
| Property B | >80 |
| Property C | <50 |
| Property D | <2 |
| Property E | >4 |

It is important to note here that the design variables were chosen to be truly independent. Thus, the component ratios were chosen to be linear and symmetrical about a central value, in contrast to the usual mixture-type designs(12,13). Further, the assumption was made that the Variable II/Variable III ratio was least likely to interact with the other variables, so it was chosen to be assigned to the three-factor interaction (assumed to be negligible). The experimental points consisted of eight cube corner points and four replicates of the center point. The YATES and REGRESSION commands in MINITAB were used to analyze the data from the measured responses, and mathematical models were developed and response surfaces plotted. The ranges of the variables and the ranges of responses are given in Table II.

Table II. Design I: $2^{4-1}$ Fractional Factorial

| Variables Studied | Range | Responses Measured | Range |
|---|---|---|---|
| # Si/Chain | 10-30 | Property A | 5-105 |
| SiUMA/Variable I | 1:3 to 3:1 | Property B | 9-122 |
| % Polar Monomer | 5-10 | Property C | 37-56 |
| Variable II/ Variable III | 1:3 to 3:1 | Property D | 0.75 - 2.6 |
|  |  | Property E | 1-5 |

MINITAB readily produces many useful manipulations of data such as were obtained in this experiment. Figure 2 shows histograms of the responses, indicating that, for the limited number of data points, the experimental values for each response approach a normal distribution. Thus, the statistical analysis was considered valid. Table III shows a copy of the computer printout of a correlation table with all the responses. Clearly, Property A and Property B are negatively correlated, as predicted, but Property B and Property E are not well correlated.

Figure 3 shows the output of the Yates analysis of Property B response, and Figure 4 shows the regression analysis. The adjusted $R^2$ is high (95%), so this model explains the data very well. Also,

```
--- HIST C7

  PROPT A

  MIDDLE OF    NUMBER OF
  INTERVAL     OBSERVATIONS
     0.           0
    10.           2     **
    20.           0
    30.           7     *******
    40.           1     *
    50.           0
    60.           0
    70.           0
    80.           0
    90.           1     *
   100.           0
   110.           1     *

--- HIST C14

  PROPT B

  MIDDLE OF    NUMBER OF
  INTERVAL     OBSERVATIONS
    10.           1     *
    20.           1     *
    30.           0
    40.           3     ***
    50.           4     ****
    60.           1     *
    70.           0
    80.           0
    90.           0
   100.           0
   110.           0
   120.           2     **

--- HIST C11

  PROPT C

  MIDDLE OF    NUMBER OF
  INTERVAL     OBSERVATIONS
    37.           1     *
    38.           0
    39.           0
    40.           0
    41.           0
    42.           1     *
    43.           0
    44.           0
    45.           2     **
    46.           0
    47.           1     *
    48.           0
    49.           1     *
    50.           1     *
    51.           0
    52.           1     *
    53.           2     **
    54.           0
    55.           1     *
    56.           1     *

--- HIST C9

  PROPT D

  MIDDLE OF    NUMBER OF
  INTERVAL     OBSERVATIONS
    0.8           1     *
    1.0           1     *
    1.2           1     *
    1.4           3     ***
    1.6           2     **
    1.8           0
    2.0           3     ***
    2.2           0
    2.4           0
    2.6           1     *

--- HIST C19

  PROPT E

  MIDDLE OF    NUMBER OF
  INTERVAL     OBSERVATIONS
     1.           2     **
     2.           2     **
     3.           2     **
     4.           3     ***
     5.           3     ***
```

Figure 2.   Histograms of responses from Design I.

YATES ANALYSIS OF VARIANCE

RESPONSE - PROPT B

```
 STANDARD   ORDER    *       RESULTS IN ABSOLUTE DESCENDING ORDER
                     *
            OBSERVED *    EFFECT    MEAN       MEAN
            RESPONSE *    DESC      EFFECT     SQUARE         F
                     *
     1      121.3000 *    B         -62.3000   7762.5801    356.91
     2       63.0000 *    A         -45.6000   4158.7201    191.21
     3       36.5000 *    AB         18.7000    699.3800     32.16
     4       17.0000 *    AC         -6.7000     89.7800      4.13
     5      121.3000 *    BC          3.1000     19.2200      0.88
     6       51.0000 *    C          -2.9000     16.8200      0.77
     7       44.1000 *    ABC        -0.7000      0.9800      0.05
     8        9.8000 *
```

AVERAGES

```
  CUBE POINTS           58.0000
  CENTER POINT          49.5250    (    4 OBSERVATIONS)
  OVERALL               55.1750
```

ERROR ESTIMATE         VARIANCE   DF

```
  FROM REPLICATION       21.7491    3   (USED IN F-CALCULATION)
```

T-VALUE FOR DIFFERENCE BETWEEN CUBE POINT AND
    CENTER POINT AVERAGE =    2.968  WITH   3  DF.
    (THIS IS A TEST STATISTIC FOR RESPONSE CURVILINEARITY)


THE 95% CONFIDENCE LIMITS FOR THE EFFECTS ARE + or -   10.4946


END OF YATES FOR PROPT B

Figure 3.  Yates ANOVA for Property B from Design I.

```
THE REGRESSION EQUATION IS
Y =    55.2 -  22.8 X1 -  31.2 X2
    -  1.45 X3 - 0.350 X4 + 9.35 X5
    -  3.35 X6 + 1.55 X7

                                  ST. DEV.     T-RATIO =
         COLUMN    COEFFICIENT    OF COEF.     COEF/S.D.
         --             55.175       2.313        23.86
    X1   #SI/CHN       -22.800       2.833        -8.05
    X2   SI/VAR I      -31.150       2.833       -11.00
    X3   %POLAR         -1.450       2.833        -0.51
    X4   VII/VIII       -0.350       2.833        -0.12
    X5   AB              9.350       2.833         3.30
    X6   AC             -3.350       2.833        -1.18
    X7   BC              1.550       2.833         0.55

THE ST. DEV. OF Y ABOUT REGRESSION LINE IS
S = 8.012
WITH ( 12- 8) =   4 DEGREES OF FREEDOM

R-SQUARED = 98.0 PERCENT
R-SQUARED = 94.6 PERCENT, ADJUSTED FOR D.F.

ANALYSIS OF VARIANCE

 DUE TO        DF        SS       MS=SS/DF
REGRESSION      7    12747.48      1821.07
RESIDUAL        4      256.78        64.20
TOTAL          11    13004.26

FURTHER ANALYSIS OF VARIANCE
SS EXPLAINED BY EACH VARIABLE WHEN ENTERED IN THE ORDER GIVEN

 DUE TO        DF        SS
REGRESSION      7    12747.48
#SI/CHN         1     4158.72
SI/VAR I        1     7762.58
%POLAR          1       16.82
VII/VIII        1        0.98
AB              1      699.38
AC              1       89.78
BC              1       19.22


DURBIN-WATSON STATISTIC = 1.18
```

Figure 4.  Regression analysis for Property B from Design I.

Table III. Correlation Table from MINITAB

|            | Property A | Property D | Property C | Property B |
|------------|-----------|-----------|-----------|-----------|
| Property D | 0.116     |           |           |           |
| Property C | -0.694    | -0.206    |           |           |
| Property B | -0.780    | -0.250    | 0.361     |           |
| Property E | -0.342    | 0.393     | -0.032    | 0.368     |

it is clear that only the #Si/chain, SiUMA/Variable I ratio, and their interaction are important variables for this response (critical $F_{1,3}$=10.13). Similar analysis of all the responses led to Table IV, which lists all the coefficients and adjusted $R^2$ for these responses. Apparently, not all the data are adequately described by the mathematical models, as shown by the low $R^2$ for some. This point will be discussed below.

One of the main reasons for deriving models of the responses is to use the models for predicting the responses in regions of the design space not actually covered by experiment. Such prediction can be done mathematically, but it is usually easier for the experimenter to look at graphical representations of the data. Again, computer software is available to plot response models over various regions of the design space (i.e., various levels of the variables).

Figures 5 and 6 show the response surfaces plotted for Property A and Property B, respectively. Note that two variables are plotted at once, with the values of the other variables fixed at levels chosen by the experimenter. The contours in the graph represent constant levels of the response. Fortunately, the computer allows rapid replotting for various levels of the fixed variables, as well as changing the identities of the fixed and floating variables, so that the entire design space can be investigated.

Inspection of Figure 5 shows that the desired level of Property A is achieved above and to the right of the "I" contour. Similar inspection of Figure 6 shows that the Property B target is below and to the left of the "E" contour. Overlaying these two graphs allows the identification of a small region of overlap of the acceptable regions: a range of variable levels which gives materials which simultaneously satisfy both property targets.

Similar overlaying of other response surface plots led to conclusions regarding the formulation variables and their effects on the properties of the copolymers. In addition, another (proprietary) computer program was used, which allowed the combination of several regression equations (for the various responses) and the calculation of variable values needed to achieve any desired combination of response values (if the models permit).

Note, however, there are two critical limitations to these "predicting" procedures. First, the mathematical models must adequately fit the data. Correlation coefficients ($R^2$), adjusted for degrees of freedom, of 0.8 or better are considered necessary for reliable prediction when using factorial designs. Second, no predictions outside the design space can be made confidently, because no data are available to warn of unexpectedly abrupt changes in direction of the response surface. The areas covered by Figures 8 and 9 officially violate this latter limitation, but because more detailed

Table IV. Models for Design I

| Variables | Property A | Property B | Property C | Property D | Property E |
|---|---|---|---|---|---|
| CONSTANT | 39.042 | 55.175 | 48.667 | 1.599 | 3.250 |
| A (#Si/Chain) | 21.470 | -22.800 | -2.130 | -0.005 | -0.875 |
| B (SiUMA/Var I) | 26.690 | -31.150 | -2.630 | 0.202 | -0.125 |
| C (% Polar Monomer) | -3.620 | -1.450 | -1.130 | 0.340 | 0.125 |
| D (Var II/Var III) | -1.000 | -0.350 | 0.370 | -0.048 | 0.375 |
| A x B | 11.700 | 9.350 | -3.380 | 0.050 | 0.875 |
| A x C | -0.053 | -3.350 | 1.630 | -0.268 | -0.125 |
| B x C | -4.000 | 1.150 | 1.620 | 0.250 | 0.125 |
| R squared, adjusted for degrees of freedom | 0.795 | 0.946 | 0.079 | 0.513 | 0.327 |

```
DESIGN I: 2(4-1) FRACTIONAL FACTORIAL

                    RESPONSE SURFACE CONTOUR PLOT
                         RESPONSE = PROPT A
VERTICAL AXIS:
 SI/VAR I
                    HORIZONTAL AXIS: #SI/CHN

           -2.000    -1.200    -0.400     0.400     1.200     2.000
           +....+....+....+....+....+....+....+....+....+....+....+
   2.0000  FF      EE      D       CC      BB      AA
   1.8750    FF      E       DD      CC      BB      AA
   1.7500      F     EE      DD      CC      BB      A
   1.6250  G     FF      EE      DD      C       B       AA
   1.5000    GG      FF      EE      D       CC      BB      AA
   1.3750      GG      FF      E       DD      CC      BB      AA
   1.2500  H     G       F       EE      DD      CC      B       A
   1.1250    H     GG      FF      EE      DD      C       BB      AA
   1.0000      HH      GG      FF      EE      D       CC      BB      AA
   0.8750        HH      GG      F       E       DD      CC      BB      AA
   0.7500  II      HH      G       FF      EE      DD      CC      B       A
   0.6250    II      H       GG      FF      EE      DD      C       BB      AA
   0.5000      I       HH      GG      FF      EE      D       CC      BB      A
   0.3750  J     II      HH      GG      F       E       DD      CC      BB
   0.2500    JJ      II      HH      G       FF      EE      DD      CC      B
   0.1250      JJ      II      H       GG      FF      EE      DD      C       BB
   0.0000        J       I       HH      GG      FF      EE      D       CC
  -0.1250          JJ      II      HH      GG      F       E       DD      CC
  -0.2500            JJ      II      HH      G       FF      EE      DD      CC
  -0.3750              JJ      II      H       GG      FF      EE      DD
  -0.5000          J       I       HH      GG      FF      E       D
  -0.6250            JJ      II      HH      GG      F       EE      DD
  -0.7500              JJ      II      HH      G       FF      EE      D
  -0.8750                JJ      I       H       GG      FF      EE
  -1.0000                  J       II      HH      GG      FF      E
  -1.1250                    JJ      II      HH      GG      F       EE
  -1.2500                      JJ      II      HH      G       FF
  -1.3750                        JJ      I       H       GG      FF
  -1.5000                          J       II      HH      GG      FF
  -1.6250                            JJ      II      HH      GG
  -1.7500                              JJ      II      H       G
  -1.8750                                JJ      I       HH      GG
  -2.0000                                  J       I       H       G
           +....+....+....+....+....+....+....+....+....+....+....+
           -2.000    -1.200    -0.400     0.400     1.200     2.000

VALUES OF CONTOUR LINES:
   A =    90.0000   B =    80.0000   C =    70.0000   D =    60.0000
   E =    50.0000   F =    40.0000   G =    30.0000   H =    20.0000
   I =    10.0000   J =     0.0000

DESCRIPTION OF PLOT CONDITIONS:
   THE RESPONSE IS      PROPT A
   HORIZONTAL AXIS IS #SI/CHN
   VERTICAL AXIS IS    SI/VAR I
   LEVELS OF OTHER INDEPENDENT VARIABLES:
      %POLAR         =    0.0000
      V II/V III     =    0.0000
      AB             =    0.0000
      AC             =    0.0000
      BC             =    0.0000
```

Figure 5. Response surface contour plot for Property A from Design I.

```
                DESIGN I: 2(4-1) FRACTIONAL FACTORIAL


                         RESPONSE SURFACE CONTOUR PLOT
                             RESPONSE = PROPT B
 VERTICAL AXIS:
 SI/VAR I
                         HORIZONTAL AXIS: #SI/CHN

            -2.000     -1.200     -0.400      0.400      1.200      2.000
            +....+....+....+....+....+....+....+....+....+....+....+....+
   2.0000   II    JJ
   1.8750    II    JJ
   1.7500   H   II    JJJ
   1.6250    HH    II    JJ
   1.5000      HH    II    JJ
   1.3750   G    HH    II     JJ
   1.2500    GGG    HH    III    JJ
   1.1250      GG    HH    II    JJ
   1.0000   FF    GG    HH    II    JJ
   0.8750    FFF    GG    HHH    II    JJ
   0.7500   E    FF    GG    HH    II    JJJ
   0.6250    EE    FF    GG    HH    II    JJ
   0.5000      EE    FF    GG    HH    II    JJ
   0.3750   DD    EEE    FF    GGG    HH    III    JJ
   0.2500    DD    EE    FF    GG    HH    II    JJ
   0.1250   C    DD    EE    FF    GG    HH    II    JJ
   0.0000    CC    DD    EE    FF    GG    HH    II    JJ
  -0.1250      CC    DDD    EE    FFF    GG    HHH    II    JJJ
  -0.2500   BB    CC    DD    EE    FF    GG    HH    II    JJ
  -0.3750    BB    CC    DD    EE    FF    GG    HH    II     JJ
  -0.5000   A    BB    CCC    DD    EEE    FF    GG    HH    II    JJ
  -0.6250    AA    BB    CC    DD    EE    FF    GGG    HH    III    J
  -0.7500      AA    BB    CC    DD    EE    FF    GG    HH    II
  -0.8750        AA    BB    CC    DD    EE    FF    GG    HH    II
  -1.0000           AA    BBB    CC    DDD    EE    FFF    GG    HHH
  -1.1250            AA    BB    CC    DD    EE    FF    GG    HH
  -1.2500              AA    BB    CC    DD    EE    FF    GG    H
  -1.3750                AAA    BB    CC    DD    EE    FF    GG
  -1.5000                  AA    BB    CCC    DD    EEE    FF    GGG
  -1.6250                    AA    BB    CC    DD    EE    FF
  -1.7500                      AA    BB    CC    DD    EE    FF
  -1.8750                        AA    BBB    CC    DDD    EE    FF
  -2.0000                           A    B    C    D    E
            +....+....+....+....+....+....+....+....+....+....+....+....+
            -2.000     -1.200     -0.400      0.400      1.200      2.000

 VALUES OF CONTOUR LINES:
     A =    120.0000   B =    110.0000   C =    100.0000   D =     90.0000
     E =     80.0000   F =     70.0000   G =     60.0000   H =     50.0000
     I =     40.0000   J =     30.0000

 DESCRIPTION OF PLOT CONDITIONS:
     THE RESPONSE IS     PROPT B
     HORIZONTAL AXIS IS  #SI/CHN
     VERTICAL AXIS IS    SI/VAR I
     LEVELS OF OTHER INDEPENDENT VARIABLES:
         %POLAR       =    0.0000
         V II/V III   =    0.0000
         AB           =    0.0000
         AC           =    0.0000
         BC           =    0.0000
```

Figure 6. Response surface contour plot for Property B from Design I.

experimentation was already planned, the risk of inaccurate predictions was considered acceptable. Note, also, that because of the uncertainty in the models, the actual area of overlap could be larger or smaller than predicted from Figures 5 and 6.

With these provisos, conclusions were derived from the results of this first experimental design. There was no difference in the effects of Variable II vs. III for any of the properties measured. Because Variable II was judged easier to work with in the lab and more stable than Variable III, Variable II was chosen as the polar monomer in future work. Both the Property C and Property D targets appeared easily attainable. As had been foreseen, Property A was mutually exclusive of both Properties B and E. However, note that adequate models were not obtained for Properties C, D, or E, so conclusions based on these models in Table IV are questionable.

In order to optimize the formulation, a different experimental design was used. Based on the results of the first design, the particular molecular weight of SiUMA was chosen which seemed to have the best chance of giving the desired balance of properties. Further, it had been established that Variable II (and not Variable III) was preferable for further work. With these variable types (i.e., not involving a component amount) eliminated, the problem was reduced to a "constrained mixture design"(12,13) involving three components: SiUMA-18, Variable I, and Variable II.

The component amounts were restricted to only certain levels, based on the results of the first design. The limits of the components, and the responses measured for the polymers prepared, are shown in Table V. This situation is nicely depicted on triangular

Table V. Design II: 3 Component Constrained Mixture Design

| Variables Studied | Range | Responses Measured | Range |
|---|---|---|---|
| SiUMA-18 | 0.3-0.45 | Property A | 5-19 |
| Variable I | 0.35-0.6 | Property B | 40-115 |
| Variable II | 0.1-0.2 | Property C | 40-51 |
| | | Property D | 1.4-3.8 |
| | | Property E | 2-4.5 |

graph paper, in which each vertex represents 100% of one of the components. Figure 7a shows the full factor space (each component from 0 to 100%), with only the cross-hatched area being covered by the formulations.

MINITAB standard regression analysis does not perform very well when the variables do not change much, because MINITAB gives error messages that certain variables are highly correlated and then refuses to include them in the analysis. In addition, when graphing response surfaces, it is difficult to see the contours within a small design space. To avoid these problems the investigational area of Figure 7a was mathematically expanded into Figure 7b. Each "pseudo" variable is linearly related to its corresponding "real" variable by the relationships:

P-SiUMA  = 4 x SiUMA - 1.2
P-Var I  = 4 x Var I - 1.4
P-Var II = 4 x Var II - 0.4

The triangle in 7b is defined as the smallest one which includes all the vertices of the hatched area in 7a. The relatively large change in the pseudo-variables allowed MINITAB to perform the regression analysis successfully. The formulations chosen for this design are shown in Figure 7b as extreme vertices, edge centroids, and an overall centroid, for a total of 12 formulations, including four replicates of the overall centroid. This choice fully and evenly covers the factor space.

For the regression analysis of a mixture design of this type, the NOCONSTANT regression command in MINITAB was used. Because of the constraint that the sum of all components must equal unity, the resultant models are in the form of Scheffe' polynomials(13), in which the constant term is included in the other coefficients. However, the calculation of correlation coefficients and F values given by MINITAB are not correct for this situation. Therefore, these values had to be calculated in a separate program. Again, the computer made these repetitive and involved calculations easily. The correct equations are shown below (13):

$$R^2_{adjusted} = 1 - \frac{SSError/(N-p)}{SSTotal/(N-1)}$$

$$F_{regression} = \frac{SSRegress/(p-1)}{SSError/(N-p)}$$

where N = number of observations, and p = number of coefficients in the mixture regression model. Figure 8 shows a typical printout of the regression analysis for Property E.

After this numerical analysis was done, another set of models was developed and conclusions were drawn from the data. These models are given in Table VI. Note that excellent models were obtained for three of the five responses, and that Properties A and E were again mutually exclusive. Figure 9 and 10 show the response surfaces for Properties A and E. Overlaying these plots shows that no combination of components was possible which allowed the simultaneous achievement of the target values of these two properties.

At this time in the investigation, the discovery was made that incorporation of a different hydrocarbon monomer at modest levels improved Property E of these copolymers. Thus, another designed experiment was necessary to determine the optimal formulation. We decided to return to the factorial design because we wanted to reexamine the effects of SiUMA chain length in combination with Variable IV. The design is set out in Table VII.

Note that, again, three different types of variables were combined: chain length, component ratio, and absolute component level. Thus, a "standard" constrained mixture design was not appropriate. In this case a full factorial, central composite design was used, with a total of 20 data points. The star points were

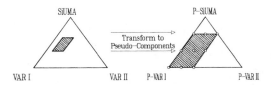

Figure 7. Constrained mixture design, showing relationship between real and pseudocomponents.

```
THE REGRESSION EQUATION IS
Y = +  2.08 X1 +  4.04 X2 +  1.93 X3
    -  0.801 X4 -  4.61 X5 +  6.61 X6
    + 14.4 X7

                                     ST. DEV.    T-RATIO =
         COLUMN      COEFFICIENT     OF COEF.    COEF/S.D.
NOCONSTANT
X1       PSIUMA         2.085          1.858       1.12
X2       PVAR I         4.0401         0.3997     10.11
X3       PVAR II        1.933          4.780       0.40
X4       AB            -0.801          4.304      -0.19
X5       AC            -4.61          10.69       -0.43
X6       BC             6.611          7.970       0.83
X7       ABC           14.42          16.40        0.88

THE ST. DEV. OF Y ABOUT REGRESSION LINE IS
S = 0.4096
WITH ( 12- 7) =    5 DEGREES OF FREEDOM

ANALYSIS OF VARIANCE

  DUE TO       DF         SS        MS=SS/DF
REGRESSION      7      152.1611     21.7373
RESIDUAL        5        0.8389      0.1678
TOTAL          12      153.0000

FURTHER ANALYSIS OF VARIANCE
SS EXPLAINED BY EACH VARIABLE WHEN ENTERED IN THE ORDER GIVEN

  DUE TO       DF         SS
REGRESSION      7      152.1611
PSIUMA          1       62.3472
PVAR I          1       81.7921
PVAR II         1        5.4857
AB              1        1.4884
AC              1        0.4075
BC              1        0.5104
ABC             1        0.1298

COLUMN     F REGR       RSQR       RSQRADJ     MSTOTAL     MSREGR     MS RESID
COUNT          1          1           1           1           1           1
ROW
   1       11.9970    0.935050    0.857109     1.17424     2.01295     0.167788

COLUMN   MS ERROR    F LK FIT
COUNT        1           1
ROW
   1     0.0625000    4.71154
```

Figure 8. NOCONSTANT regression analysis for Property E from Design II.

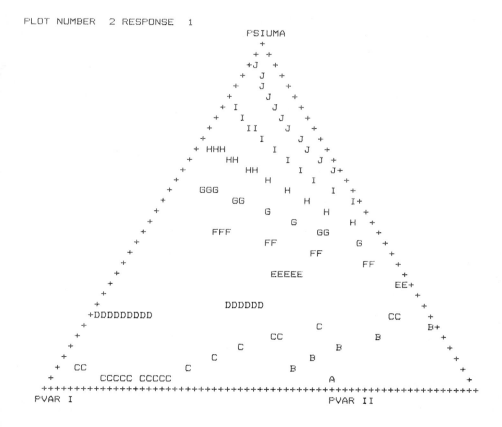

Figure 9. Response surface contour plot for Property A from Design II.

54  COMPUTER APPLICATIONS IN THE POLYMER LABORATORY

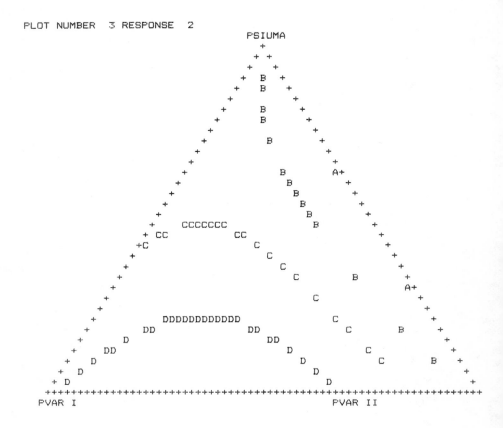

Figure 10. Response surface contour plot for Property E from Design II.

Table VI. Models for Design II

| Variables | Property A | Property B | Property C |
|---|---|---|---|
| A (P-SiUMA-18) | 20.48 | 43.46 | 62.86 |
| B (P-Variable I) | 4.99 | 116.7 | 51.53 |
| C (P-Variable II) | -7.24 | 205.15 | 52.01 |
| A x B | -2.03 | 5.13 | -39.53 |
| A x C | 39.72 | -271.55 | -55.07 |
| B x C | 20.91 | -220.43 | -32.21 |
| A x B x C | -80.13 | 161.86 | 44.87 |
| R squared, adj. for degrees of freedom | 0.975 | 0.885 | 0.393 |

| Variables | Property D | Property E |
|---|---|---|
| A (P-SiUMA-18) | 1.036 | 2.085 |
| B (P-Variable I) | 2.617 | 4.04 |
| C (P-Variable II) | -0.003 | 1.933 |
| A x B | -0.951 | -0.801 |
| A x C | 13.35 | -4.61 |
| B x C | 6.8 | 6.611 |
| A x B x C | -7.05 | 14.42 |
| R squared, adj. for degrees of freedom | 0.435 | 0.857 |

Table VII. Design III: $2^3$ FULL FACTORIAL CENTRAL COMPOSITE DESIGN

| VARIABLES STUDIED | RANGE | RESPONSES MEASURED | RANGE |
|---|---|---|---|
| #Si/Chain | 16-29 | Property A | 1-34 |
| SiUMA/Variable I | 35/65 to 55/45 | Property B | 40-115 |
| | | Property C | 50-57 |
| % Variable IV | 2-8 | Property D | 2.4-3.3 |

necessary because we observed that second order terms in the regression equations were required to fit the data. The star points were measured in a separate block, along with two center point replicates. The "block effect" was not explicitly included in the Yates analysis, even though it theoretically should have been.

The same analysis techniques were used in this third design as were used in the first design. The final models and $R^2$ values are shown in Table VIII. Note that models for Properties C, D, and E are not given. Measurements for the first two responses were not taken on the star point formulations.

Property E is missing because of the inability to develop a laboratory test to quantify the response adequately. This situation is an example of two things. First, there is always a problem of

Table VIII. Models from Design III

| Variables | Coefficients | |
|---|---|---|
| | Property A | Property B |
| Constant | 19.214 | 68.603 |
| A (#Si/Chain) | 2.7 | -2.3 |
| B (SiUMA/Variable I) | 9.724 | -24.525 |
| C (% Variable IV) | -1.815 | 2.363 |
| A x B | -1.875 | 1.125 |
| A x C | -4.125 | 6.625 |
| B x C | -1.375 | 0.375 |
| A x B x C | -3.875 | 4.125 |
| ASQR | -3.119 | 9.4 |
| BSQR | 0.519 | 2.035 |
| CSQR | -0.012 | 2.212 |
| R squared, adjusted for degrees of freedom | 0.922 | 0.883 |

devising laboratory tests which accurately predict reality. Second, results of analysis and computer printouts can look quite impressive; but computers cannot think. The expertise of the experimenter is always necessary to interpret these results properly. Although computers made the analysis of the design data convenient, they could not tell that the test being used to generate the Property E data was itself faulty.

Two significant conclusions from this last design were: 1) the laboratory test for Property E was not a good predictor of actual performance, and 2) a small amount of Variable IV was important to get a material with good Property E. Based on the results of this last design, a formulation was identified which was predicted to meet all the conflicting property goals. The properties of the final, optimized formulation are given in Table IX. Comparison of the measured properties with the targets shows that we were indeed able to reach those targets.

Table IX. Final Properties

| Property | Target | Measured* |
|---|---|---|
| Property A | >15 | 14 + 1 |
| Property B | >80 | 100 + 15 |
| Property C | <50 | 50 + 2 |
| Property D | <2 | 2 + 0.2 |
| Property E | >4 | No test |

*Values are average with standard deviation from at least three replicates.

## Conclusion

In conclusion, two types of statistically designed experiments were used in this investigation to reach an understanding of the range of properties available from copolymers using the new SiUMA oligomer. Full and fractional factorial experiments allowed the combination of different variable types including component identity and level. A constrained mixture design with pseudo-components was used to investigate relative component levels when the component identities were fixed. Various computer programs were used to design and analyze the results of these experiments. The use of these techniques rapidly and efficiently led to the desired understanding and to the identification of a formulation which met a particular set of properties targets. The use of computers was essential in allowing rapid analysis, both statistical and graphical, of the large amount of data generated in these experiments.

## Literature Cited

1. Merker, R. L. and Noll, J. E., J. Org. Chem. 21, 1537 (1956).
2. Gaylord, N. G., U.S. Patent 3,808,179 (1974).
3. Eckberg, R. P., U.S. Patent 4,348,454 (1982).
4. Taiyo Ink Seizo K. K. Shikoku Fine Chemicals K. K., Jpn. Patent JP 57 59.966 (1982).
5. Makovicka, J., Mares, J., and Kolman, B., Czech. Patent CS 213,155 (1984).
6. Williams, T. R., to be published in J. Applied Polymer Sci.
7. McGrath, J. E., Riffle, J. S., Yilgor, I., Banthia, A. K., and Sormani, P., Amer. Chem. Soc., Prepr. Org. Coating Plast. Chem. 46, 693 (1982).
8. Lukevics, E., Belyakova, Z. V., Pomerantseva, M. G., and Vorokov, M. G., Organometal. Chem. Rev. 5, 1 (1977).
9. Refojo, M. F., et al., Contact and Intraocular Lens Medical J. 3, 27 (1977).
10. "Submerged Wetting Angle", Contact Lens Manufacturers Assoc., Chicago, adopted February, 1981.
11. Box, G. E., Hunter, W. G., and Hunter, J. S., "Statistics for Experimenters", John Wiley & Sons, New York, 1978.
12. Snee, R. D., Chemtech 702 (1979).
13. Cornell, J. A., "Experiments with Mixtures: Designs, Models, and the Analysis of Mixture Data", John Wiley & Sons, New York, 1981.

RECEIVED November 14, 1985

# 7

# Analysis and Optimization of Constrained Mixture-Design Formulations

**Stephen E. Krampe**

**Medical–Surgical Division, 3M, St. Paul, MN 55144**

> Techniques are presented for the analysis of mixture design responses and for the optimization of formulations comprised of components constrained by lower and upper limits. The analysis of component effects by the use of tricoordinate contour plots is enhanced by an algorithm which determines the feasible experimental region defined by the component limits. The direct optimization of a single response formulation modelled by either a normal or pseudocomponent equation is accomplished by the incorporation of the component constraints in the Complex algorithm. Multiresponse optimization to achieve a "balanced" set of property values is possible by the combination of response desirability factors and the Complex algorithm. Examples from the literature are analyzed to demonstrate the utility of these techniques.

A considerable amount of statistical effort has been devoted to the design of mixture experiments, the proper selection of regression models to analyze the designs, and the determination of the individual component effects on the response. Such research is warranted due to the unique dependencies inherent in mixture designs, the restrictions on the component levels which the experimenter typically imposes on the system, and the considerable number of applications utilizing formulations. This paper takes the end result of the mixture design, the equations describing the responses in terms of the component levels, and presents three techniques which enhance the information available from the design. These techniques assume that good statistical practice has been used for the establishment of the mixture design and the development of the response equation.

## Boundary Region Determination on Tricoordinate Contour Plots

Unlike conventional experimental designs which have independent variables, mixture designs possess variables which are interdependent in that the summation of the q component proportions must be unity. Typically, the individual component levels are restricted by lower ($a_i$) and upper ($b_i$) constraints imposed on the system by physical or chemical limitations of the formulation or by the selection of the level values by the formulator. These constraints are represented as:

$$\sum_{i=1}^{q} x_i = 1$$

$$0 \leq a_i \leq x_i \leq b_i \leq 1$$

Any component level change must be compensated by changes of the remaining component levels to maintain the unity constraint. To describe the relationships between the response and the component levels, polynomial models of special forms are fit to the data. The Scheffé model (1), expressed in quadratic canonical form as

$$E(Y) = \sum_{i=1}^{q} \beta_i X_i + \sum_{1 \leq i < j}^{q} \beta_{ij} X_i X_j$$

is one form used to fit mixture response data although alternative model forms (2-4) may also be used. Note that the Scheffé model excludes the constant term due to the unity constraint imposed on the system. The significance of the model and the component effects can then be determined with a modified analysis of variance (5) and by calculation of partial and total effects (6). Although these analyses provide useful information, the formulator still faces the complex task of exploring the entire experimental space defined by the nonlinear blending relationships of the equation to find the composition yielding the optimum response.

Graphical techniques, particularly tricoordinate contour plots, aid mixture analysis by giving a visualization of the component effects using response contours. Unfortunately, only three components can be plotted against the response while the remaining components must be kept at fixed levels. In general, the lower and upper limits of the three components and the remaining fixed values define a compositional region which satisfies the summation to unity criterion. The determination of this feasible region, existing as a point, a line, or as a region of up to six sides, is important in that response contours outside of this region represent extrapolations of the mixture model. One additional technique to enhance the model fitting and graphical analysis involves the use of pseudocomponents (7,8). This technique involves the linear transformations of all component values according to the following equations:

$$L = 1 - \sum_{i=1}^{q} a_i \qquad x_i' = \sum_{i=1}^{q} (x_i - a_i)/L$$

Computationally, the use of pseudocomponents improves the conditioning of the numerical procedures in fitting the mixture model. Graphically, the expansion of the feasible region and the rescaling of the plot axes allow a better visualization of the response contours.

The perceptional advantages of response contours in illustrating nonlinear blending behavior and the additional information of the experimental boundary locations were incorporated into a generalized algorithm which determines the feasible region on a tricoordinate plot for a normal or pseudocomponent mixture having any number of constrained components.

Given the components to be plotted and the fixed values of the remaining components, the algorithm first tests the existence of a feasible region by:

$$\sum_i a_i + \sum_{j \neq i} (\text{fixed values})_j < 1$$

$$\sum_i b_i + \sum_{j \neq i} (\text{fixed values})_j > 1$$

where in each equation the first summation includes only those components being plotted and the second summation includes only the remaining components. This existence test is necessary to insure that the experimenter's choice of fixed values will result in a feasible region for the plotted components. The feasible region is then calculated as that minimal area bounded by the intersections of the component limits. Six computations involving multiple comparisons are necessary to calculate the maximum possible number of intersection points of a three component plot. Equivalencies of intersection point values are indicative of a region having fewer than six sides. In this manner, feasible regions having any number of sides up to the maximum number of six may be easily determined and plotted on the tricoordinate graph. To complete the enhancement of the tricoordinate plots, the algorithm rescales the plot axes by linear transformations of the minimum and maximum feasible limits of the plotted components.

<u>Example: Feasible Region Determination and Rescaling</u>. McLean and Anderson (<u>9</u>) described a mixture experiment in which magnesium ($X_1$), sodium nitrate ($X_2$), strontium nitrate ($X_3$), and binder ($X_4$) were combined and ignited to produce flares varying in intensity. The four components had the following ranges:

$$.40 \leq X_1 \leq .60$$

$$.10 \leq X_2 \leq .50$$

$$.10 \leq X_3 \leq .50$$

$$.03 \leq X_4 \leq .08$$

Gorman (7) transformed the mixture variables to pseudocomponents to develop the following equation:

$$\hat{y}(\underline{X}^{\prime}) = -208.2X_1^{\prime} + 124.3X_2^{\prime} + 62.0X_3^{\prime} + 2908.0X_4^{\prime}$$

$$+1136.2X_1^{\prime}X_2^{\prime} + 1105.6X_1^{\prime}X_3^{\prime} - 904.5X_1^{\prime}X_4^{\prime}$$

$$+439.9X_2^{\prime}X_3^{\prime} - 2324.5X_2^{\prime}X_4^{\prime} - 2342.2X_3^{\prime}X_4^{\prime}$$

Figure 1 represents the contour plot of intensity values when the value of $X_4$ = .05. The pseudocomponent equation is used for the contour determinations, but the component values on the axes were scaled to their normal values for ease of interpretation. Given the original constraint limits and the fixed value, the algorithm computes the feasible region as the area within the bold lines on the plot. Additional tricoordinate plots at other values of $X_4$ could be analyzed to find the maximum flare intensity and the corresponding optimal formulation. For compositions comprised of a large number of components, this method of optimization is very inefficient due to the difficulty in discerning the interdependent relationships. A numerical optimization technique will be discussed in the next section to aid formulation analysis.

Single Response Mixture Optimization

The objectives of a formulator in performing a mixture design are to not only determine the component effects and blending relationships but also optimize the component levels to achieve a maximum or minimum response of a measured property. Unfortunately, the mixture design literature is sparse in references to mixture optimization. McLean and Anderson (9) in the classic flare example attempted to use Lagrange multipliers to maximize the equation describing the intensity of an ignited flare composition but obtained erroneous results. However, a secondary technique which was not discussed did produce the optimum.

Other established techniques to aid in the analysis of mixture models include the use of gradients to measure the rate of response change (10), graphical analysis of the response change versus individual component changes (11), and the determination of component effects within constrained regions (6). Each of these techniques, while very useful in the interpretation of component effects do not lend themselves to the determination of the optimum composition, the most important point in the formulation space.

Constrained, multivariate optimization algorithms, such as the general use of Lagrange multipliers, the incorporation of constraints by the use of penalty functions (12), gradient projection techniques (12), and various incremental step searches accelerating in both direction and distance within constrained regions (14,15) are useful for particular classes of optimization but possess certain drawbacks when applied to mixture designs. The use of Lagrange multipliers or the determination of the differential equations for the gradient techniques is generally difficult for the formulator. A major problem in their use is variable interdependence due to the summation to unity constraint. This unity constraint could potentially be incorporated into some of these algorithms as penalty functions but the determination of new points during the optimization process would still be difficult since component level compensation would have to be included. The procedure to achieve component compensation is critical for a mixture optimization algorithm. Piepel (6) discussed the method of compensation given a change in any one component but to incorporate this univariate approach within an optimization algorithm in which all levels are being varied would not be effective.

Efforts directed toward the evaluation and modification of optimization algorithms for mixtures have resulted in the identification of an algorithm which avoids the separate calculation of compensatory changes. This algorithm was developed by Box (16) as the Complex method and was based on the earlier work of Nelder and Mead (17) for unconstrained optimization. The general procedure involves the progression to an optimum by means of function evaluations at simplex vertices, rejection of the vertex yielding the worst value, and reflection of the worst point through the simplex centroid to obtain a new vertex. Evaluation of the function at the new point then results in either an acceleration step in distance beyond the new vertex in the same direction or a contraction step back toward the simplex centroid.

Several features make this algorithm particularly attractive for the optimization of a formulation response. The algorithm requires only the input of the lower and upper limits of the individual components and the equation describing the response. Both of these must be expressed in either normal or pseudocomponent form. A randomization procedure generates the initial simplex within the individual component constraints by:

$$X(I) = XLOW(I) + R[XHIGH(I) - XLOW(I)]$$

where
$X(I)$ = simplex value of component I
$XLOW(I)$ = low limit of component I
$XHIGH(I)$ = high limit of component I
$R$ = computer generated random number, $0 \leq R \leq 1$

A new random number is used for each calculation resulting in a component level within its own compositional limits. The final component level is then calculated as one minus the summation of the previously determined values. If the final component is not within its own constraint limits, the process is reinitiated with a new calculation of the first component value. Each set of feasible formulation levels generated in this manner corresponds to one vertex point. The Box recommendation of using twice the number of vertices as components was followed for the formulation optimization.

The greatest advantage of this technique is apparent during sequential generation of new simplex vertices in proceeding to the optimum. If all the initial simplex vertices comply with the overall mixture constraint of having the summation of component values sum to unity (inherent in the initial generation of the vertex points), then the new vertices generated by the reflection and contraction process will yield new component values which also comply with the unitary constraint. This point being dependent on the simplex centroid for the direction determination. This compensation of component changes, a shortfall in other algorithms, is inherently accomplished by the Complex algorithm.

After each reflection or expansion step, the component levels of the new vertex point are tested to assure compliance within the individual lower and upper constraints. If during the search procedure a constraint limit is violated by a particular component, a correction factor is calculated to force the component value to remain in the feasible space at the boundary limit value. For a constraint being violated by component I, this correction factor is computed as:

$$a = (XLOW(I) - XCEN(I))/(XCEN(I) - X(I))$$

where
$\quad$ a = correction factor
$\quad$ XLOW(I) = low limit component I
$\quad$ XCEN(I) = simplex centroid value of I
$\quad$ X(I) = value of I prior to step

The correction factor is then applied to each of the remaining component values (J) to determine the new boundary vertex composition. The new values (XNEW) of the components then become:

$$XNEW(J) = XCEN(J) + a[XCEN(J) - X(J)]$$
$$XNEW(I) = XLOW(I)$$
$$\text{where } I \neq J$$

If the high limit is violated, the value of XHIGH(I) replaces the XLOW(I) term. With the new vertex point on the constraint boundary, the algorithm continues the reflection and contraction process until all simplex vertices are within a preset convergence criterion. The composition at convergence may represent a local or global optimum due to the nonlinear behavior of the response and should be tested with new starting vertices to assure that the global optimum of the response has been achieved. The maximization or minimization of a mixture model is performed equally well by this algorithm.

<u>Example: Optimization of the Four Component Flare Mixture</u>. The formulation of the previously discussed flare example was optimized by the Complex algorithm to yield the maximum intensity. The McLean and Anderson mixture equation (9), fit using normal component values and constraints, produced an optimum formulation at $X_1 = .5232$, $X_2 = .2299$, $X_3 = .1669$, and $X_4 = .0800$. The Gorman pseudocomponent equation for the same mixture data (7), constrained using pseudocomponent values, produced a virtually identical optimum at $X_1 = .5233$, $X_2 = .2298$, $X_3 = .1669$, and $X_4 = .0800$ when expressed in normal form. The inherent compensatory characteristic of the Complex algorithm makes this technique applicable for the formulation optimization of mixture designs modelled in either normal or pseudocomponent form.

<u>Example: Four Component Pesticide Example</u>. Cornell (2) presented the analysis of a four component pesticide mixture containing Vendex ($X_1$), Omite ($X_2$), Kelthane ($X_3$), and Dibrom ($X_4$) to control mites on plants. The response, the average percentage of mites seven days

after treatment relative to the initial number, was fit by the following model with $S^2 = 220.4$:

$$\hat{y}(\underline{X}) = 1.8X_1 + 25.4X_2 + 28.6X_3 + 38.5X_4 - 34.8X_1X_2 - 48.4X_1X_3$$
$$+ 34.2X_1X_4 - 94.4X_2X_3 + 21.8X_2X_4 - 238.5X_1X_3X_4 - 40.8X_2X_3X_4$$
$$- 1439.2X_1X_2X_3X_4 \quad \text{with } 0 \leq X_i \leq 1 \text{ and } \sum_{i=1}^{4} X_i = 1$$

Cornell discussed the significance of the component effects and nonlinear blending behavior by examination of the coefficients of the fitted model and by contour plots. No optimal formulation, however, was determined by this analysis. A minimization of the response by the Complex algorithm yielded the optimum response of -2.204 achieved at:

| Factor | Optimum Value |
|---|---|
| $X_1$ | .5041 |
| $X_2$ | .3354 |
| $X_3$ | .0000 |
| $X_4$ | .1605 |

The negative response at the optimal composition can be attributed to the inherent error in the estimation of the model coefficients. These results indicate that the best composition does not include the Kelthane ($X_3$) component. A contour plot of the response at the optimal Kelthane value ($X_3 = 0$), Figure 2, shows the sensitivity of the response to the remaining components. Additional sensitivity analyses may be performed by plotting the response versus individual component changes as shown by Snee and Marquardt (11) or Piepel (6) using the optimum composition as the reference point.

Example: Optimization of an Eleven Component Glass Formulation. Piepel (6) discussed the generation and analysis of a mixture design consisting of eleven oxides used to prepare glasses for waste vitrification. Although many responses must be considered for the end use of this composition, the intent of Piepel's study was to minimize the response of leachability subject to the compositional constraints of:

$$.41 \leq X_1 \leq .60 \quad\quad 0 \leq X_5 \leq .09 \quad\quad 0 \leq X_9 \leq .035$$
$$.055 \leq X_2 \leq .15 \quad\quad .09 \leq X_6 \leq .17 \quad\quad 0 \leq X_{10} \leq .035$$
$$0 \leq X_3 \leq .16 \quad\quad 0 \leq X_7 \leq .065 \quad\quad 0 \leq X_{11} \leq .035$$
$$0 \leq X_4 \leq .14 \quad\quad 0 \leq X_8 \leq .080$$

$$\text{and the additional constraints } \sum_{i=1}^{11} X_i = 1$$
$$.54 \leq X_1 + X_2 + X_3 \leq .80$$
$$.13 \leq X_4 + X_5 + X_6 \leq .35$$

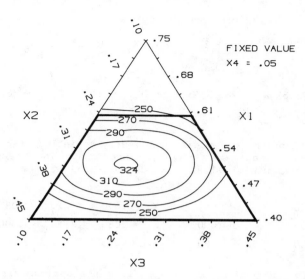

Figure 1. Flare intensity contour responses at a fixed binder level of .05.

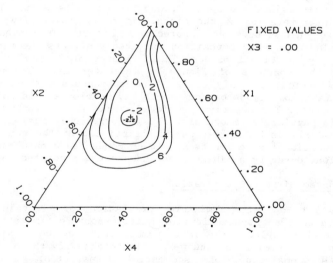

Figure 2. Contour sensitivity analysis illustrating the effect of compositional changes on the response around the optimal formulation.

The fitted model for leachability was:

$$\hat{y}(\underline{X}) = -6.2X_1 + 3.2X_2 + 5.3X_3 - 0.5X_4 + 3.0X_5 + 10.X_6 - 3.3X_7$$

$$-4.6X_8 - 0.2X_9 + 0.0X_{10} + 18.6X_{11} + 15.1X_6X_8 - 112.4X_3X_{11}$$

$$-78.8X_7X_9 - 78.3X_5X_6 + 88.0X_7X_{11} + 13.4X_1X_5 - 76.9X_6X_{10}$$

$$+34.7X_5X_9 \qquad \text{with } R^2 = .984$$

The Complex algorithm was modified slightly to include the two extra summation constraints as penalty functions. The response of leachability was minimized at a value of -3.53 with the following formulation:

| | | | |
|---|---|---|---|
| $X_1 = .600$ | $X_4 = .045$ | $X_7 = .060$ | $X_{10} = .035$ |
| $X_2 = .055$ | $X_5 = .000$ | $X_8 = .080$ | $X_{11} = .000$ |
| $X_3 = .000$ | $X_6 = .090$ | $X_9 = .035$ | |

The effects of compositional changes from the optimum are shown by plotting the contour responses of any three components while fixing the levels of the remaining components at the optimal formulation. The technique for determination of the feasible region and the plot scaling, as previously discussed, now proves to be extremely useful for the interpretation of the plots. Figure 3, a plot of $X_7$, $X_9$, and $X_{11}$, shows that $X_9$ does not appreciably affect the response while $X_{11}$ if included in the composition, adversely affects the response in an inverse proportional manner. The influence of $X_6$, $X_7$, and $X_8$, as shown by the contours of Figure 4, indicates that component $X_6$ is quite influential if only these three components are permitted to vary. Other combinations of components may be plotted and analyzed in similar manner. The negative predicted value of leachability at the optimum is physically impossible. This occurrence is again attributed to the model's lack of response description over the compositional ranges. Additional experiments should be conducted around the optimal formulation to verify a minimal response at the predicted optimum.

## Multiresponse Mixture Optimization

The ability to optimize a mixture formulation for a single response also represents an opportunity for multiresponse optimization. Cohon and Marks (18) reviewed the large number of techniques available for multiresponse optimization and were able to classify these into three categories consisting of sequential generation, noniterative ordering, and progressive preferencing. The choice of an algorithm for multiresponse optimization is dependent not only on the type and amount of available information but also on philosophical considerations such as the definition of a multiresponse optimum and the criteria used to justify tradeoffs between responses. For the mixture design case, two techniques appear to be applicable.

Khuri and Conlon (19) utilized a noniterative ordering variation of goal programming to minimize a distance criterion between the optimum of individual responses of a conventional experimental

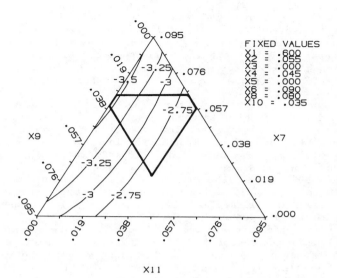

Figure 3. Effects of components $X_7$, $X_9$, and $X_{11}$ on the leachability response around the optimal formulation of the glass example.

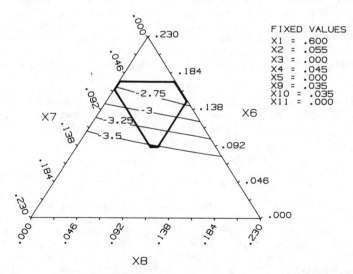

Figure 4. Effects of components $X_6$, $X_7$, and $X_8$ around the optimal formulation.

design. A preliminary analysis is conducted to remove a response linearly dependent on other responses. An extensive random search over the feasible region using a technique by Price (20) determines the general regions where the optimum (local or global) may exist. Within each individual region, a constrained gradient algorithm (21) searches for the optimal point. A comparison of values then provides the global optimum.

Accountability of the variances and covariances of the responses makes this optimization procedure particularly noteworthy. From the formulator's viewpoint, the distance criterion could lead to an unacceptable optimum if the formulation levels at the optimum produce response values in an undesirable property range. Khuri and Conlon mentioned the possible use of this procedure for multiresponse mixture optimization although no elaboration or examples were given. The reliance of this method on the gradient projection technique could present difficulties with component level compensation if applied to formulations.

A more subjective approach to the multiresponse optimization of conventional experimental designs was outlined by Derringer and Such (22). This sequential generation technique weights the responses by means of desirability factors to reduce the multivariate problem to a univariate one which could then be solved by iterative optimization techniques. The use of desirability factors permits the formulator to input the range of property values considered acceptable for each response. The optimization procedure then attempts to determine an optimal point within the acceptable limits of all responses.

Acceptable limits may differ considerably for the measured property values, however. For a majority of cases, responses can be classified as having either a maximum acceptable value, a minimum acceptable value, or a most desirable value within an acceptable range. For each response, these cases may be incorporated into a desirability factor having the range $0 \leq d_i \leq 1$ with $d = 0$ denoting an unacceptable response value and $d_i = 1$ representing totally acceptable behavior. Intermediate values are determined by equations (22) based on the acceptable limits, the desired response value, and the type of response behavior desired by the formulator.

The individual desirability factors are then combined into an overall desirability function by the geometric mean:

$$D = (d_1 \times d_2 \times d_3 \times \ldots \times d_k)^{1/k}$$

where k = number of responses

The consolidation of individual values into an overall value is possible by a number of other methods. The geometric mean calculation has a particularly distinguishing characteristic that if any individual desirability factor is zero, indicating that a response is outside of its acceptable limits, the overall desirability is zero. An optimization procedure to maximize the overall desirability function will therefore seek to remain within the acceptable ranges of all responses even though individual maximization, minimization, or an acceptable range of response values are required. By these criteria, the values maximizing the overall desirability function represent the best "balanced" conditions subject to the initial set of responses and the selected response limits.

Derringer and Suich used a Hooke-Jeeves optimization (23) for the analysis of multiresponse situations from conventional experimental designs. For the multiresponse optimization of formulations, the concept of desirability factors to incorporate response information was combined with the Complex algorithm for the optimization of a formulation. Several minor modifications to the Complex algorithm were necessary to accommodate the desirability function as a response. To perform as a multiresponse optimization procedure, the algorithm must be initiated with a simplex consisting of formulation points within the constraints of all individual component limits and response acceptability limits so as to produce a nonzero overall desirability value. As discussed previously, the randomized generation of component levels creates the initial simplex of formulations to comply with the compositional constraints. However, these formulations may produce an overall desirability value of zero due to individual responses having unacceptable predicted values. Provision in the algorithm is made for an automatic reinitiation of the randomized component level procedure until all simplex starting compositions produce a positive overall desirability value. Conceivably, the acceptable response limits may have totally exclusive compositional regions in which case no starting compositions will be found. To account for this situation, a maximum number of randomized compositional searches must be included in the algorithm. The formulator must then change the accceptability ranges of the responses to avoid the totally exclusive response behavior. Analysis of the response values from the mixture design should provide the formulator a good guide in the establishment of nonexclusive response limits within the experimental objectives. If regression equations adequately describe the responses and the desired response limits are not compositionally exclusive, then the maximum number of components and responses to be included in the multiresponse optimization is limited only to the size of the experiment which the formulator wishes to conduct.

Example: Three Component Multiresponse Optimization. An adhesive composition consisting of three components was under developmental evaluation. Twelve formulations were prepared consisting of compositions within the following pseudocomponent ranges:

| Variable | | Constraints | |
|---|---|---|---|
| | | Lower | Upper |
| Acrylate Monomer | $X_1$ | 0.0 | 1.0 |
| Hydrophilic Monomer | $X_2$ | 0.0 | 1.0 |
| Shear Monomer | $X_3$ | 0.0 | 0.4 |

Regression analysis was conducted on measured values for three responses to yield the following equations:

Initial Adhesion = $86.6X_1 + 82.8X_2 - 53.5X_3 - 45.0X_1X_2$
(s.e. = 2.2)
$+ 109.9X_1X_3 + 134.2X_2X_3 - 32.0X_1X_2X_3$

Aged Adhesion = $214.2X_1 + 197.3X_2 + 114.1X_3 + 12.3X_1X_2$
(s.e. = 4.5)
$- 250.6X_1X_3 + 1.2X_2X_3 - 656.8X_1X_2X_3$

Residue = $1.96X_1 + 1.59X_2 - .86X_3 + .27X_1X_2 + 4.4X_1X_3$
(s.e. = .05)
$+ 5.18X_2X_3 - 1.47X_1X_2X_3$

Product performance dictated that the initial adhesion values remain within lower and upper limits while the aged adhesion and residue values have minimum and maximum limits respectively. With these considerations in mind, the experimenter established the following ranges on each response:

|  | Limit Values | | |
| --- | --- | --- | --- |
|  | Lower | Desired | Upper |
| Initial Adhesion | 60 | 70 | 80 |
| Aged Adhesion | 150 minimum | | |
| Residue | | | 1.7 maximum |

The desirability factors using these values were incorporated into the Complex algorithm and optimized to give a formulation representing the best "balance" of properties given the compositional and response constraints. The optimal formulation had an overall desirability value of .27 and had the composition with $X_1$ = .332, $X_2$ = .519, and $X_3$ = .149. Without sensitivity analysis to show the effects of component changes, the optimal formulation is limited in information. Figure 5 shows the effect of component changes on the overall desirability value. A nonzero overall desirability value, indicative of the overlap of acceptable response limits, is achieved only within limited compositional ranges. The contours show that the shear monomer has the smallest latitude for change before unacceptable performance results. Figures 6-8 show the individual contour plots of the equations for the three responses. The shear monomer level appears to be bounded by the aged adhesion constraint at the higher level and the residue constraint at the lower level.

In this example with only three components, the optimum could have been determined by simply overlaying the individual response contour plots. This approach would be difficult, if not impossible, if the formulation would have many responses or contain four or more components. By contrast, the combination of the desirability function and the Complex algorithm permits an optimization of a multiresponse formulation having many constrained components in addition to providing the basis for sensitivity analysis.

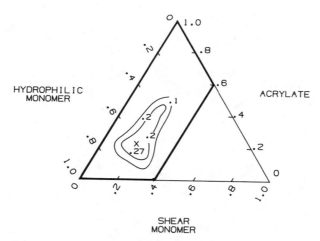

Figure 5. Multiresponse sensitivity analysis of the overall desirability value showing the optimum and the compositional region complying with all response limits.

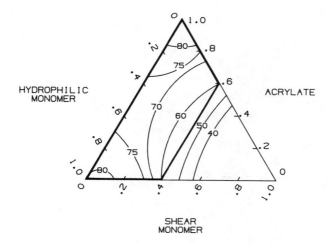

Figure 6. Initial adhesion response contours for the multiresponse optimization example.

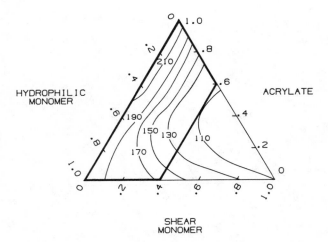

Figure 7. Aged adhesion response contours.

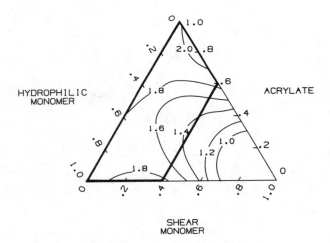

Figure 8. Residue response contours.

## Summary

Three techniques for the analysis of mixture designs have been presented. Visualization of the contour response trends on a tricoordinate plot allows the formulator to quickly determine the effects of the components within a bordered experimental region determined by the compositional constraints. The optimal formulation with respect to a single response can be determined by a modification of the Complex algorithm. This algorithm effectively compensates the component levels to maintain the unity constraint during the optimization procedure and produces an optimal formulation within the given limits. The multiresponse optimization of a formulation to achieve a "balance" of properties is possible by the combination of desirability factors which impart individual response information and the Complex algorithm. Contour sensitivity analyses may then be conducted around the optimal formulation to determine compositional influence.

## Literature Cited

1. Scheffé, H. J. R. Statistic. Soc. B. 1958, 20, 344.
2. Cornell, J.A. "Experiments with Mixtures"; John Wiley & Sons: New York, 1981; pp. 20-25, 212-256.
3. Snee, R. D. Technometrics 1973, 517.
4. Draper, N.R.; St. John, R. C. Technometrics 1977, 19, 37.
5. Marquardt, D. W.; Snee, R. D. Technometrics 1974, 16, 533.
6. Piepel, G. F. Technometrics 1982, 24, 29.
7. Gorman, J. W. J. Qual. Technol. 1970, 2, 186.
8. Crosier, R. B. Technometrics 1984, 26, 209.
9. McLean, R. A.; Anderson, V. L. Technometrics 1966, 8, 447.
10. Cornell, J. A.; Ott, L. Technometrics 1975, 17, 409.
11. Snee, R. D.; Marquardt, D. W. Technometrics 1976, 18, 19.
12. Fiacco, A. V.; McCormick, G. P. "Nonlinear Sequential Unconstrained Minimization Techniques"; John Wiley & Sons: New York, 1968.
13. Rosen, J. B. J. Soc. Indust. Appl. Math. 1960, 8, 181.
14. Rosenbrock, H. H. Computer J. 1960, 3, 175.
15. Rosenbrock, H. H.; Storey, C. "Computational Techniques for Chemical Engineers"; Pergamon Press: New York, 1966.
16. Box, M. J. Computer J. 1965, 8, 42.
17. Neider, J. A.; Mead, R. Computer J. 1964, 7, 308.
18. Cohon, J. L.; Marks, D. H. Water Resources Research 1975, 11, 208.
19. Khuri, A. I.; Conlon, M. Technometrics 1981, 23, 363.
20. Price, W. L. Computer J. 1977, 20, 367.
21. Carroll, C. W. Operations Res. 1961, 9, 169.
22. Derringer, G.; Suich, R. J. Qual. Technol. 1980, 12, 214.
23. Hooke, R.; Jeeves, T. A. J. Assoc. Comp. Mach. 1962, 8, 212.

RECEIVED December 24, 1985

# INSTRUMENT AUTOMATION
# FOR POLYMER CHARACTERIZATION

# 8

## Advantages of Interfacing a Viscoelastic Device with a High-Speed and -Capacity Computer
### and an Advanced Statistical-Graphics Software Package

**Stephen Havriliak, Jr.**

Bristol Research Laboratories, Rohm and Haas Company, Bristol, PA 19007

There are many advantages to the polymer scientist for constructing such interfaces. First of all, a high storage computer allows one to store large amounts of data in a form that makes data storage and retrieval, as well as comparisons, quite simple and efficient. Secondly, complicated calculations, involving experimental design, error estimation or conventional polymer science theories of one form or another can readily be carried out. For example, a replication study could be carried out and variance or standard deviations be represented in terms of measured quantities. Viscoelastic data could then be presented in a form of experimental quantities along with confidence limits. When polymer systems are compared, one can estimate the differences in behavior relative to the estimated error (signal/noise ratio) and select the optimum conditions for comparison.

We have found situations where comparisons can be made in regions where the experimental error may be high but the differences in observed behavior more than compensate for the error. Smoothing functions that reduce the noise in the system can also be used so that the useful range of the instrument can be extended several fold. Blend calculations using any of the well known expressions can be used to estimate the properties of blends from the properties of the pure components and their volume fraction. The results of such calculations can then be compared to experimental values and the magnitude of the experimental error to determine whether the differences are meaningful or not. This paper reviews the advantages to the polymer scientist when such an interface is used.

### EXPERIMENTAL

All the viscoelastic measurements were carried out in the Rheometrics Dynamic Spectrometer RDS-770 at a frequency of 1Hz, a strain of 0.1%, and a temperature range of $-140°$ to $140°C$ incremented every 2 degrees. The Texas Instrument Terminal Silent 700 was tapped to provide a hookup to an IBM 308X main frame computer located some miles away. The output of the Rheometrics unit was converted to a data file to be used in conjuction with SAS (1). All statistical manipulations, software developments, and the necessary graphics that are reported here were carried out with the aid of SAS.

Setting the incremental temperature change to $2°C$ resulted in an average $2°C$ change but there were many places where the change was 1 or 3 degrees. Since these changes occurred randomly and variance calculations or other comparisons must be performed at the same temperature for each material, a small program was written to interpolate the data to 1 degree increments. The SAS data files consisted of real (GSP) and imaginary (GDP) parts of the modulus, the loss tangent (tandel), temperature (T) and the sample identification (sampno).

The three polymers that were chosen for study, e.g. PMMA (2), EPDM (3), and Hytrel (4), were selected because they represent a wide range of viscoelastic materials. These materials were processed into plaques. The plaques were annealed at $125°C$ between highly polished chrome plated flat plates and cooled slowly to minimize the effects of residual stresses. Viscoelastic measurements were made under conditions cited above on two test specimens that were cut from adjacent locations in the plaque to minimize possible preparation variations. The test specimen size in all cases was 2 x 0.5 x 0.125 in. Blends of PMMA/Hytrel =3/1 were also prepared by extrusion blending in a 1 in. Killion extruder. The pellets were injection molded into plaques. These plaques were annealed at 125°C for 3 minutes, and then cooled to room temperature slowly to minimize residual stresses.

## RESULTS

### A. ESTIMATES OF EXPERIMENTAL ERROR

Variance (5,6), related to the scatter in replicated experiments, was calculated using SAS'S PROC MEANS for storage modulus, loss modulus and loss tangent for each temperature and material, e.g., PMMA, EPDM AND HYTREL. Various relationships were examined to select the proper meter for relating variance to the magnitude of measurement, e.g. linear, semi-log and log-log. The most satisfactory meter that was found was the log-log relationship. SAS's PROC STEPWISE was used to construct a model for representing the variance in terms of the experimental variables, such as the magnitude of the measurements, their cross-terms and temperature. Temperature was found to contribute only slightly in these relationships. SAS's PROC GLM was used to estimate the contribution of polymer type. This was found to be only a weak correlation and need not be included. Table I summarizes the results and analysis and the assumptions that were used to construct the final algebraic equation to represent the variance. 95% confidence limits were then calculated from the expressions in Table II, which were derived from SAS's PROC GLM, the simplified model and an estimate of the residuals. It should be pointed out that the numerical values calculated from these expressions should be used carefully (see Discussion) since they include important assumptions in their derivation. It is expected that other laboratories may derive similar relationships but with somewhat different numerical values.

An example of the utility of these expressions is given in Figure 1 where the log (real modulus) is plotted against temperature for two test specimens of EPDM. The first test specimen is represented by o's, while the second specimen is represented by the dashed lines, which are 95% confidence limits calculated from the expressions in Table II and it viscoelastic properties. The measurements were made three weeks apart.

Table I. Summary of R-Square Values for the Various Variance Models

| MODEL TYPE | TANDEL* | GSP* | GDP* |
|---|---|---|---|
| 1. SAS's PROC STEPWISE, no polymer type | .45 | .90 | .75 |
| 2. SAS's PROC GLM, polymer type is a class variable | .41 | .93 | .82 |
| 3. Simplified model, see Table II | .41 | .88 | .74 |

*TANDEL is the loss tangent, GSP is the storage modulus and GDP is the loss modulus.

Table II. Numerical Expressions Used to Represent the 95% Confidence Limits of the Viscoelastic Quantities

1. Log(VDT)=-3.13902+2.1813 Log(TD)-0.144921 Log(GSP)+0.01117 PROD

    SDTD=(VTD/0.41)$^{1/2}$   and   VTD=10*EXP(Log(VTD))

2. Log(VGSP)=-2.0235+0.02906 Log(TD)+2.4905 Log(GSP)+0.6803 PROD

    SDGSP=(VGSP/0.88)$^{1/2}$   and   VGSP=10EXP(Log(VGSP))

3. Log(VGDP)=2.40928+2.6814 Log(GDP)-0.6744 Log(GSP)-0.38506 PROD1

    SDGDP=(VGDP/0.74)$^{1/2}$   and   VGDP=10EXP(Log(VGDP))

Note that GSP and GDP have been reduced by 10exp9 with PROD=Log(TANDEL) Log(GSP) and PROD1=Log(GSP) Log(GDP)

Note that VTD=variance of Loss Tangent, and that SDTD is the standard deviation of Loss Tangent with similar definitions for GSP (G' or real modulus) and GDP (G" or loss modulus).

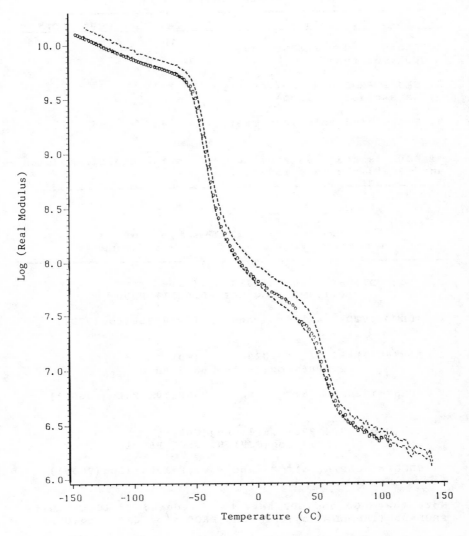

Figure 1. A plot of the log (storage modulus) for EPDM with temperature. The o's represent the first run while the confidence limits of the second run are represented by the dashed lines.

Note how the confidence band changes with changing temperature. As an aside, it should be pointed out that at a temperature of about $0°C$, the machine torque is below the recommended range but the transducer is still responding to the signal.

## B. SIGNAL/NOISE RATIO

Suppose one wanted to compare the behavior of two polymers and their blends. Let us define the signal as the difference between the logarithims of the viscoelastic quantities and the noise as the error calculated for a specific set of viscoelastic properties associated with a specific composition. The signal to noise ratio would have the appearance of the three curves shown in Figure 2 for a PMMA/Hytrel blend =3/1. Selection of the optimum conditions for comparison is apparent in that figure. Emphasis should be placed at those temperatures with high signal/noise ratios.

## C. SMOOTHING FUNCTIONS

The properties of variance noted earlier allow us to treat the scatter in the results as noise. A number of smoothing functions exist for this purpose. One such method that is particularly suited since the incremental temperature change has been kept constant is that due to Savitsky and Golay (7). This method is a rigorous application of the method of least squares to those relationships that can be represented by polynomials and for experiments that are uniformly spaced in the independent variable. In the present case the temperature range was uniformly spaced at 1 degree intervals and the viscoelastic functions can be represented by polynomials no greater than degree 3. The results of smoothing the Hytrel modulus temperature curve is shown in Figure 3.

Another advantage of the Savitsky-Golay method is that derivatives of these functions can also be determined from the method of least squares. This method can be used to determine alpha-peak temperatures automatically since the first derivative changes sign at the peak temperature. The advantage of smoothing is that the number of extraneous peaks due to noise has been minimized.

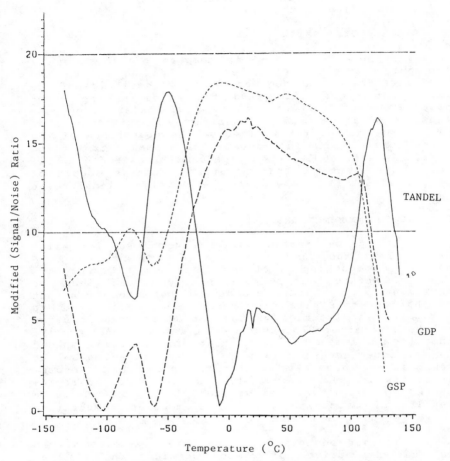

Figure 2. Plot of the signal/noise ratio as defined in the text with temperature for a polyblend of PMMA and HYTREL.

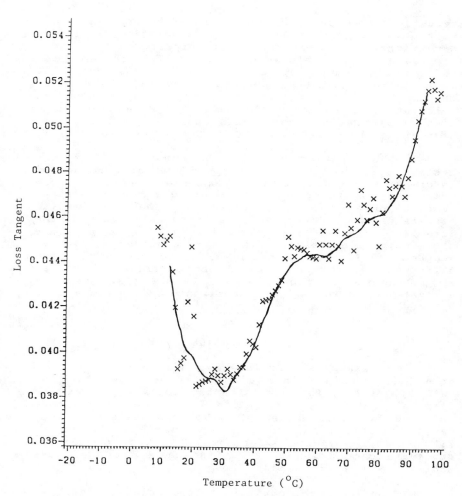

Figure 3. Plot of the loss tangent with temperature for Hytrel in the temperature range of 0 to 140 Deg. The x's represent the experimental values, while the solid line represents the results of smoothing using the Savitzky-Golay technique.

## D. VISCOELASTIC PROPERTIES OF BLENDS

The SAS data files that were created as a results of the operations noted earlier have the important property that all three viscoelatic properties have either been measured or interpolated to the same temperature. Therefore one can merge data sets for different polymers by temperature (a SAS data set manipulation) and then perform blend calculations on the data quite simply in SAS. For example, one can calculate the expected viscoelastic properties of a blend from the pure components and their volume fraction using the equations of Uemura and Takaynagi (8).

These results can then be compared to experimental values (at the same temperature) in a number of informative ways. First we can plot the calculated values as a function of temperature and represent results as a line, see Figure 4. The experimental results can be represented as an error band vs temperature plot. Real differences are readily apparent since they must lie outside the 95% confidence limits. Another way to represent the difference is to plot the difference between calculated and experimental values as a function of temperature. In the same graph an estimate and plot of the experimental errors can also be made, see Figure 5.

## DISCUSSION

The examples described here, as well as others, have been combined into a software package which we call POLYREOM. The user of this routine has at his disposal a Menu so that he can select any of the comparisons described above. In this way, quite complicated calculations can be carried out simply by anyone with minimal computer experience. In fact the entire procedure can be run on a routine basis from the time that a measurement is made.

Some important conditions concerning the estimation of error should be pointed out. First, modulus measurements of rectangular bars are made in torsion and the calculations contain assumptions that may depend on geometry. How this influences error, particularly at low torque levels is not known. Second, the strains were kept constant at 0.1%; other strains might not yield the same results. On the other hand one would expect an inverse proportionality to exist between the magnitudes of error and strain. Thirdly, these errors were estimated for a frequency of 1Hz.

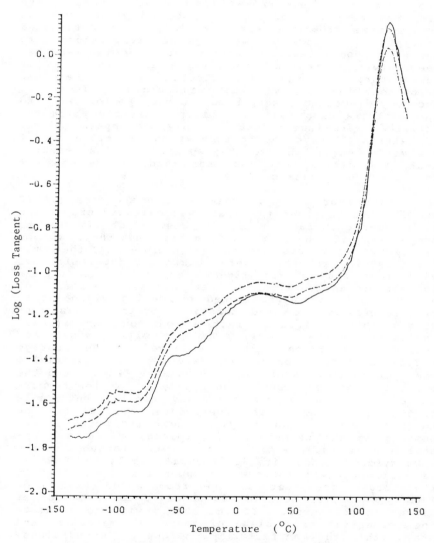

Figure 4. Plot of the loss tangent with temperature for PMMA/HYTREL blend=3/1. The solid line represents the calculated values while the dashed lines represent the 95% confidence limits for the experimental values.

Finally, the contribution to error, that is specific to this laboratory, especially when compared to other laboratories, is not known. On the other hand, there are some very interesting and noteworthy observations to be made from the error study. First, repeated attempts to include polymer type and temperature into the error model failed. This observation implies that temperature control of the device is independent of temperature and that any fluctuation is the same throughout the interval that was used in this study. Secondly, repeated attempts to include sample type into the analysis also failed, implying that the estimated variances were independent not only of temperature but of material. This result suggests very strongly that the variances are due to the magnitude of the measurement, the device that was used, and our own set of laboratory circumstances.

There are many polymer science theories (9, 10 and 11) that attempt to relate viscoelastic properties to polymer structure or, in the case of blends, to the properties of the pure components and their volume fraction. Implicit in these theories is that the viscoelastic properties were obtained under controlled conditions and in the linear region. Linearity in polymers is an idealization that may never be realized, and may only be approximated by making viscoelastic measurements at low and constant strain. This constraint places serious restrictions on experimental equipment as the following example will demonstrate. Consider the temperature dependence of the storage modulus as a function of temperature for two materials, where at room temperature one is glassy (PMMA), and the other is leathery (EPDM), see Figure 6. At low temperatures, the moduli are not only similar but quite high. As a result of experimental conditions, the torque, a machine quantity, is near its upper limit. However as the temperature is raised, the curves behave quite differently. At $0°C$ the torque for the EPDM system is near its lower limit, while for PMMA it is still near the instruments upper limit. Eventually the torque goes out of the recommended range for EPDM. A similar behavior is observed for PMMA but at higher temperatures. If one should wish to estimate the properties of a blend of these two materials and stay within the manufacturer's torque recommendations, then the temperature range is severely limited because any interactions with the alpha and beta processes of PMMA cannot be studied. Changing the dimensions of the specimens, the transducer, or the level of strain are not acceptable alternatives to minimize the effects of a limited torque range.

One way to minimize these effects is derived from the results of this work. For example, the

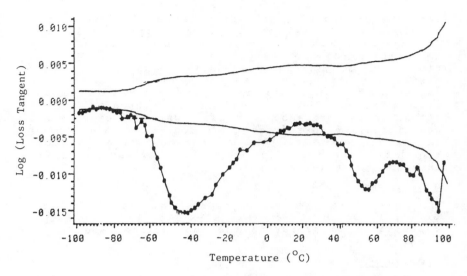

Figure 5. Plot of the loss tangent difference with temperature for data in Figure 4. The circles represent the experimental values, while the line represents the 95% confidence limits for the experimental values.

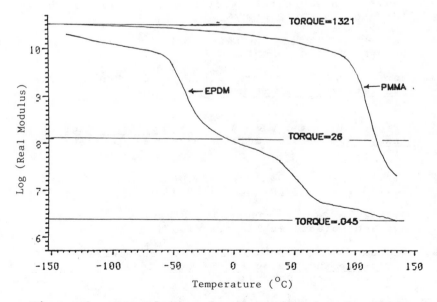

Figure 6. Dependence of storage modulus for PMMA and EPDM on temperature at a frequency of 1Hz and a strain of 0.1%.

increase in experimental error is not as important as is the signal/noise ratio. A knowledge of this ratio is important because optimal conditions (in this case temperature) can be selected for comparison. In addition, use of the Savitzky-Golay technique to reduce error is justified because the error was found to be essentially noise, and not a low value machine cutoff. The problem becomes one of searching for the signal in the presence of noise. Using 26 experimental quantities, which represents 26 degrees of measurement, is not unreasonable. One would expect an error reduction of 5, thus leading to, or an extension of, the instruments range by 5 fold.

Acknowledgments

This writer wishes to express his appreciation to Dr. E. P. Dougherty and Mr. F. Miller for many helpful discussions in preparing the subroutines used in this work.

Literature Cited

1. SAS Institute, SAS Circle, Box 8000, Carry, North Carolina 27511.

2. The PMMA used in this work is Plexiglas V(811), a product of the Rohm and Haas Co.

3. The EPDM used in this work is Nordel 2722, manufactured by the Du Pont Co.

4. The Hytrel used in this work is Hytrel-4056, a product of the DuPont Co.

5. Box G.E.P., Hunter W.G., and Hunter J.S., Statistics for Experimenters, John Wiley and Sons (1978).

6. Draper N. and Smith H., Applied Regression Analysis, Second Edition, John Wiley and Sons (1981).

7. Savitzky A. and Golay M.J.E., Analytical Chemistry, Vol. 36, No. 8, pp. 1622-1639 (1964).

8. Bucknall C.B., Toughened Plastics, Chapter 5, Applied Science Publishers (1977).

9. McCrum N. G., Read B. E., and Williams G., Anelastic and Dielectric Effects in Polymer Solids, John Wiley and Sons (1967).

10. Ferry J. D., Viscoelastic Properties of Polymers, Second Edition, John Wiley and Sons (1982).

11. Mansfield M. L., J of Poly. Sci., Polymer Phys Div., Vol 21, 787-806 (1983).

RECEIVED December 24, 1985

# 9

# Analysis of Isochronal Mechanical Relaxation Scans

Richard H. Boyd

Department of Materials Science and Engineering and Department of Chemical Engineering, University of Utah, Salt Lake City, UT 84112

> A method is described for fitting the Cole-Cole phenomenological equation to isochronal mechanical relaxation scans. The basic parameters in the equation are the unrelaxed and relaxed moduli, a width parameter and the central relaxation time. The first three are given linear temperature coefficients and the latter can have WLF or Arrhenius behavior. A set of these parameters is determined for each relaxation in the specimen by means of nonlinear least squares optimization of the fit of the equation to the data. An interactive front-end is present in the fitting routine to aid in initial parameter estimation for the iterative fitting process. The use of the determined parameters in assisting in the interpretation of relaxation processes is discussed.

It has become fairly common to characterize polymeric materials by means of computer automated dynamic mechanical measurements. The measurements are often made at one or a limited number of frequencies but at a large number of temperatures. The data taken is then displayed isochronally as E' and E" as a function of temperature at one or a few frequencies. A little reflection indicates that the isochronal mode not only is a convenience but actually is a necessity in dealing with polymeric relaxations. That is, the relaxations are so broad in the frequency or time domain that even a very wide ranging (in frequency or time) isothermal measurement covers a small part of the relaxation. The glass transition region in amorphous polymers is considered to be broad. However, sub-glass relaxations and relaxations in semi-crystalline polymers (1) are much broader still. A five or six decade isothermal time or frequency range gives only a small slice of the total relaxation region in the latter relaxations. Unfortunately, it has been tradtional that, once the isochronal scans are constructed, further analysis and interpretation is usually qualitative. Further phenomenological data analysis can be implemented however. This materially aids in resolution of partially merged relaxation processes and in better quantifying and explaining the differences between various specimens. This analysis can be carried out in much the same spirit as traditional analysis of isothermal data. In fact, a single isochronal scan will do and it carries almost as much information as several scans at one or two decades in time or frequency. Several applications of this kind of analysis of recently been made (**1,2**). However. the methodology has not been described in detail and in the belief that such analysis should be generally useful the opportunity to do this is taken here.

## Rationale

It is appropriate to focus on some general parameters that could characterize a relaxation and to see how these are reflected in the experimental data. These parameters would include an unrelaxed, low temperature, high frequency modulus, $E_U$, and a relaxed, high temperature, low frequency modulus, $E_R$. The difference between these is a measure of the intensity of the process and in fact is often referred to as "relaxation strength." The time- temperature location of the relaxation can be thought of as set by a temperature dependent central relaxation time, $\tau_0$. The temperature dependence of the latter is described by an activation energy, $\Delta H^*$ (and activation entropy). Finally, the detailed appearance of the process can be thought of directly in terms of breadth and shape (symmetry) on a log frequency plot (and parameters that would set these) or more indirectly in terms of the breadth and shape of a relaxation time distribution.

Isochronal temperature scans reflect or contain information about the relaxation parameters. It is apparent (Figure 1) that the relaxed and unrelaxed moduli, $E_R$, $E_U$ are approximated by the high and low temperature extremes of the E' vs. temperature curve. It is well-known that the area of a plot of loss versus 1/T yields activation energy (for Arrhenius temperature dependence and a temperature independent distribution of relaxation times) (**3**). That is,

$$<1/\Delta H^*> = A''/2\pi R (E_U - E_R) \qquad (1)$$

where $A'' = \int E'' d\ln 1/T$ (the brackets on $1/\Delta H^*$ indicate a weighted average over the various relaxation times). Thus, for a given value of $E_U - E_R$, a sharp peak (with small area) implies high activation energy and a broad one the opposite. As mentioned, the sharpness of an isochronal scan peak is determined also by the width of the relaxation time distribution, a narrow distribution contributing toward a narrow peak in a temperature scan and vice-versa. Under favorable circumstances, this type of broadening can be distinguished from activation energy effects for the following reason (what constitutes unfavorable circumstances will be commented on later). Variations in peak width with activation energy take place at constant peak height (Figure 1), whereas variations in peak width due to relaxation time distribution changes are accompanied by peak height changes (i.e., a broad peak has a lower peak height than a sharp one, see Figure 2).

## Phenomenological description

Having suggested the connections between relaxation descriptors and the data it is now important to realize that here is sufficient information in isochronal scans that, with numerical analysis now readily carried out by computer, detailed parameters that describe relaxation can be determined jointly. Analysis is most conveniently carried out with the aid of a parameterized empirical phenomenological function. The method as implemented by us uses for each relaxation peak a Cole- Cole -like function (**4**) to represent the complex modulus,

$$E^*(i\omega) = (E_U - E_R)(i\omega\tau_0)^\alpha / (1 + (i\omega\tau_0)^\alpha) \qquad (2)$$

where $\omega = 2\pi \bullet$ frequency. The choice of modulus, as opposed to compliance, $J^*(i\omega)$, to parameterize is arbitrary and is based on the observation that data is more often reported as the former in the literature. The function in equation (2) differs from the conventional Cole-Cole one by being modified to represent (mechanical) relaxation rather than (dielectric) retardation. The explicit parameters in equation (1) consist of the central relaxation time, $\tau_0$, a width parameter, $\alpha$ ($0<\alpha\leq1$, small $\alpha$ = broad peak, $\alpha\rightarrow1$ = narrow, single relaxation time peak), and $E_U$ and $E_R$ for each process present. The equation does not allow for skewing of the loss peak in log frequency. In its most general form, the parameterized function should be the Havriliak-Negami equation (**5**) which does allow this behavior. However frequency domain dielectric studies of sub-glass processes and the glass-rubber relaxation in semi-crystalline polymers indicate that these peaks are symmetric (**1**). As the examples we have treated largely belong to this class we have made the simplification of eliminating the skewness parameter. It would have to be included in studies of the glass-rubber relaxation in wholly amorphous polymers (**1, 5**).

The explicit parameters are, in turn, temperature dependent. The limiting moduli and the width parameter are taken (for simplicity) to have linear temperature behavior,

$$E_R = E^0_R + S_R (T-T_0) \qquad (3)$$

$$E_U = E^0_U + S_U (T-T_0) \qquad (4)$$

$$\alpha = \alpha_0 + S_\alpha (T-T_0) \tag{5}$$

where the S's are linear temperature coefficients (the bold-face parameters are those chosen for determination by optimization of the fit of the function to the data). In the case of sub-glass relaxations the following modification of the width parameter equation was found to be important.,

$$\alpha = \alpha_0 + [\alpha' \Delta T + (\alpha'^2 \Delta T^2 + k)^{1/2}]/2 \tag{6}$$

$$\Delta T = (T-T_0)$$

This equation gives linear dependence of $\alpha$ on T at high temperature (with slope $\alpha'$) but a constant $\alpha_0$ is approached at low temperature (see Figure 3). The constant, k, controls the smoothing between the asymptotes. The central relaxation time is given WLF dependence (or Arrhenius with $T_\infty = 0$),

$$\log \tau_0 = A/(T-T_\infty) + B \tag{7}$$

(the activation energy in this formulation is given by $\Delta H^* = 2.303 R A T^2/(T-T_\infty)^2$).

These parameters are determined by non-linear least-squares optimization of the fit of the function to both the experimental storage and loss moduli curves. As emphasized, the two determiners of temperature-scan peak width referred to above (i.e., in terms of equation (2), activation energy $\Delta H^*$ of $\tau_0$ and $\alpha$ ) have features that allow distinquishing betweenthem in the optimization. Increasing values of $\Delta H^*$ sharpen the peak but do not increase $E''_{max}$ (at $T_{max}$). Increasing values of $\alpha$ sharpen the peak by increasing $E''_{max}$ and lowering values away from the maximum. That is, at fixed $E_U - E_R$, changes in $\Delta H^*$ change the area under the loss curve (vs. 1/T) keeping $E''_{max}$ constant but changes in $\alpha$ change the width keeping the area constant. The relaxation strength, $E_U - E_R$, is rather sensitive to the experimental E' curve and simultaneous fitting of both E' and E" is essential in ensuring good parameterization.

## Parameter determination

**Method.** Let there be a separate Cole-Cole function (equation (2)) for each relaxation process ($\alpha, \beta, \gamma$) found, or,

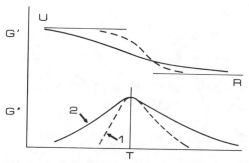

Figure 1. Effect of activation energy on width of relaxation process ($\Delta H^*_1 > \Delta H^*_2$).

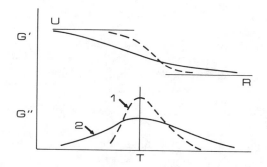

Figure 2. Effect of $\alpha$ parameter (equation (2)) on width of relaxation process ($\alpha_1 > \alpha_2$).

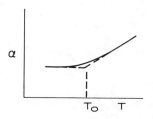

Figure 3. Temperature dependence of $\alpha$ width parameter used for sub-glass processes (equation (6)).

$$E^* = E^*(\alpha) + E^*(\beta) + E^*(\gamma) + E_R \qquad (8)$$

where $E^*$ is the total or complete complex modulus and $E_R$ is the relaxed modulus above the highest temperature process ($\alpha$). Let $E^*(i)$ be the value of equation (8) at the temperature and frequency of the experimental point, i, that has measured values $E'_i$, $E''_i$ associated with it. Then an object function for minimization can be written as,

$$F = \Sigma w'_i (E'(i) - E'_i)^2 + \Sigma w''_i (E''(i) - E''_i)^2 \qquad (9)$$

where the $w_i$'s are weights assigned to the points. The object function written this way ignores the connection between G', G" occasioned by a Kramers-Kronig-like relation. The minimization is conveniently linearized via Newton-Raphson iteration by writing,

$$F = F_0 + F' \Delta P + 1/2\, \Delta P^T F'' \Delta P + \cdots \qquad (10)$$

where **P** is a column vector containing the parameters being optimized (the bold-face constants in equations (3-7)), **F'** is the vector of first derivatives of F with respect to each parameter, $p_k$, that is, it contains the elements,

$$\partial F/\partial p_k = 2\Sigma w'_i (E'(i) - E'_i)\, \partial E'(i)/\partial p_k + 2\Sigma w''_i (E''(i) - E''_i)\, \partial E''(i)/\partial p_k \qquad (11)$$

and **F''** is the matrix of second derivatives and has elements,

$$\partial^2 F/\partial p_k \partial p_l = 2\Sigma w'_i\, \partial E'(i)/\partial p_k\, \partial E'/\partial p_l + \Sigma w'_i (E'(i) - E'_i)\, \partial^2 E'/\partial p_k \partial p_l$$

$$+ \text{similar } E'' \text{ terms} \qquad (12)$$

Updates to the parameters are found from $\delta F = 0$ as the solution to the linear equations,

$$F'' \Delta P = -F' \qquad (13)$$

and $P(\text{new}) = P(\text{old}) + \Delta P$.

**Implementation.** In our implementation, the derivatives $\partial E^*(i)/\partial p_k$ in equations (11) and (12) are calculated analytically from equations (2-7). In equation (12), the terms in $\partial^2 E^*/\partial p_k \partial p_l$ are neglected in comparison to terms in the product of first derivatives. By using complex declaration of appropriate variables in FORTRAN all rationalization of complex expressions is carried out by the compiler.

Since the fitting process is an iterative one, an initial guess for the parameters must be obtained. This is often the most troublesome and/or inconvenient aspect of any optmization procedure. We have developed an interactive front-end for our fitting routine that is both fairly convenient and effective in this context. All initial parameters are generated from responses to prompts. First the data is displayed (in our case, for device independence, on a conventional line printer daisy-chained to the CRT terminal). From this display, esitmates of $E_U$ and $E_R$ at a given temperature are entered along with estimates of the associated slopes. Following this, an estimate of the width parameter ($\alpha$) and it's temperature coefficient are asked for. Typical values are usually sufficient Then $T_{max}$, $f_{max}$ for the E" peak are entered along with a typical activation energy for the kind of process encountered. This latter value need not be at all precise. Then a value for $T_\infty$ (equation (7)) is asked for. It is not an adjusted parameter. For WLF processes, a value corresponding to a temperature just below the start of the E" peak is satisfactory. For an Arrhenius process 0 is entered. The entry process is repeated for each relaxation present. An editor is present in the front-end so that errors in entries can be corrected or revisions made before iteration begins. In Figure 4, the Initial parameter estimate construction from a typical data display is shown. The data is for a "6-6" linear aliphatic polyester and was taken on the torsion pendulum in our laboratory (6). In Figure 5, the fit resulting from the initial parameter estimates (listed in Table I) is shown. Then the iteration process is started.

Table I. Estimates of Parameters for Start of Iterative Fitting

| Process | $E_R$ | $S_R$ | $\alpha$ | $S_\alpha$ | $\Delta H^*$ | $T_{max}$ | $T_\infty$ |
|---|---|---|---|---|---|---|---|
| unrelaxed | 0.22 | −0.001 | | | | | |
| $\gamma$ [a] | 0.04 | -0.0006 | 0.20 | 0.001 | (10.0) | 140.0 | 0.0 |
| $\beta$ [b] | -0.30 | -0.0035 | 0.20 | 0.001 | 60.0 | 220.0 | 180.0 |

(a) $T_0$ in equations (3,4) is 140°K, $\alpha$ is from equation (6) with $T_0$ = 110°, act. energy constrained.

(b) $T_0$ in equations 3-5 is 220

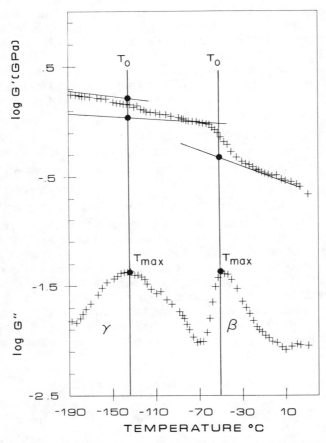

Figure 4. Construction of initial paramter estimates from data display.

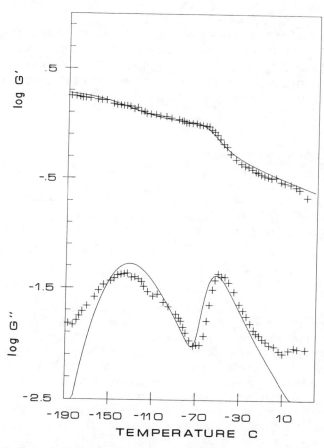

Figure 5. Calculated curves from initial parameter estimates of Figure 4.

The user has the choice of displaying the fit from each iteration or simply looking at the residuals and the changes in the parameters. Return to the initial parameters and entry into the editor can be made even after the iterative process has begun so that an obviously bad choice of initial parameters can be revised without starting execution all over again. In Figure 6, the fit after a converged sequence of 9 iterations is displayed. The converged parameters are given in Table II. The weights, $w_i$, in equation (9) are normally assigned as 1. An exception might be, for example, at the high temperature extreme where melting or extreme softening renders the data atypical or unreliable and zero weights can be assigned to those points.

Table II. Converged parameters

| Process | $E_R$ | $S_R$ | $\alpha$ | $S_\alpha$ | A | B | $T_\infty$ |
|---|---|---|---|---|---|---|---|
| unrelaxed | 0.314 | 0.0009 | | | | | |
| $\gamma$ | 0.051 | -0.0011 | 0.086 | 0.0013 | (2185.) | -19.66 | 0.0 |
| $\beta$ | -0.342 | -0.0042 | 0.216 | -0.0012 | 582.9 | -13.97 | 180.0 |

It is apparent that there are a considerable number of parameters to be determined. According to equation (8) and equations (2-7) there are 6N+2 parameters where N is the number of relaxations present (it is not 8N because the relaxed modulus of one process is equal to the unrelaxed modulus of the next process in a sequence). In practice, it is found that with the large number of experimental points available in a scan (typically 50-100) the determinaton usually proceeds satisfactorily. However, in common with many statistical fitting situations, it can happen that parameter determination is not unique. Our experience has shown that problems can arise when the relaxation strength is small or when only part of a peak is recorded. The problem with small relaxaton strength is associated with equation (1) where it is seen that the activation energy is related to the ratio of peak area and relaxation strength $E_U$- $E_R$. When the process is quite weak the experimental E' curve no longer serves to define the strength well and activation energy is thus poorly defined. Multiple frequency scans would be especially useful in such cases in helping to define the activation energy. However in their absence, problems like these can usually be handled by means of constraints. The poorly behaving parameter can be

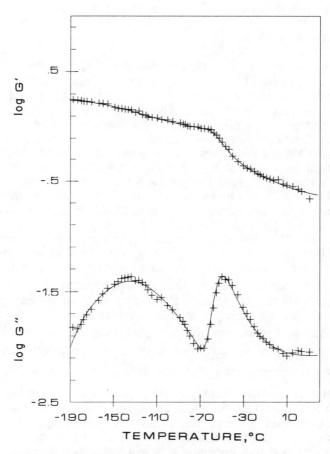

Figure 6. Calculated curves from converged parameters.

constrained to a reasonable value. Constraints on parameters are easily incorporated into equation (13) by means of Lagrange multipliers. In the data entry portion of the fitting, the question is asked if any parameters are to be constrained (and the appropriate constraint is then set up based on the response). Return to the initial parameters and addition of new constraints can be made at any point in the iterative process.

## Results and discussion

Now that the determination of parameters that describe the relaxation processes has been developed, it is appropriate to discuss some uses to which the parameters can be put in interpretation of the relaxation processes. An important area, one that provided the simulus to develop the above methodology, concerns morphological assignment of relaxation peaks in semi-crystalline polymers. Probably much of what arises here is apropriate for phase separated systems in general. It would be important if it were valid to assign a given loss peak to either the amorphous or crystalline phase. For example, one of the peaks should be the glass- rubber relaxation in the residual amorphous fraction. Confirmation of this would include appropriate variation of relaxation descriptors with degree of crystallinity. One might expect the loss peak height (a quantity directly available without phenomenological analysis) to decrease with increasing crystallinity. This turns out to be a frustrating exercise. The result depends not on the data but on what way it is displayed. The same loss data can be plotted as $E''$, $J''$ (loss compliance) or $\tan\delta$. In illustration of this, consider isotactic polypropylene. The dynamic shear modulus data of Passaglia and Martin (7) for several isotactic polypropylene samples is shown in Figure 7. Notice that for the $\beta$ relaxation, the one postulated as the glass- rubber relaxation, $\tan\delta_{max}$ and $J''_{max}$, although not well resolved, vary in the expected way with crystallinity. However $G''_{max}$, which is well resolved, varies almost not at all with crystallinity. An argument can be made that $\tan\delta$ is the safest indicator (1,8 and the reasons why the various measures of peak height differ can be explained (in terms of the composite behavior of the system) (1). However the question of the effect of poor resolution of the peak remains. A better procedure is to plot the relaxation strength ($E_U - E_R$ or $\log E_U - \log E_R$) of the process directly. This requires the phenomenological analysis. We have fit equation (2) to this data by the optimizing procedure above. Parenthetically it is apppropriate to comment that in the parameter optimization that the activation energy of the $\gamma$ process was constrained. All of the other parameters were free to vary. The $\gamma$ constraint was necessary because of the small relaxation strength and in the $\alpha$ one because not all of the peak is recorded. The curves shown in Figure 7 are the result of the fitting. When relaxation strength is plotted versus

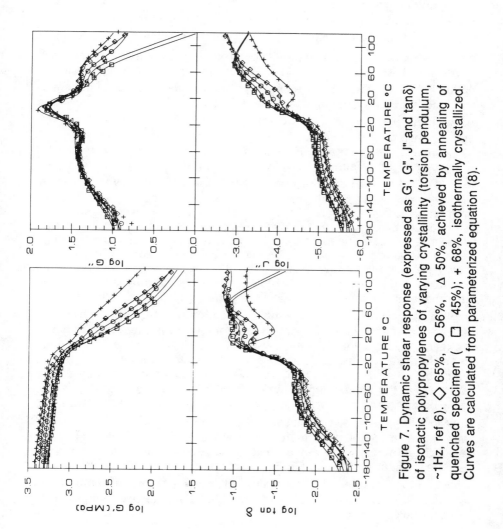

Figure 7. Dynamic shear response (expressed as G', G", J" and tanδ) of isotactic polypropylenes of varying crystallinity (torsion pendulum, ~1Hz, ref 6). ◇ 65%, ○ 56%, △ 50%, achieved by annealing of quenched specimen ( □ 45%); + 68%, isothermally crystallized. Curves are calculated from parameterized equation (8).

crystallinity (Figure 8) it becomes obvious that the β process is decreasing in strength with increasing crystallinity as is consistent with its interpretation as the glass-rubber relaxation in the amorphous fraction.

Other results are that the strength of the γ process is too small to reach a strong conclusion about phase origin from this decriptor. However tan $\delta_{max}$ is consistent with amorphous origin. Very interestingly, the α process is seen from the relaxation strength plot to be of amorphous phase origin. It is true that there is other strong evidence that the underlying molecular motion in the α process involves the crystal phase but the actual softening is transferred to the amorphous fraction (1). This seems to be generally true of mechanical α processes in crystalline polymers (where fundamental crystal phase origin is involved).

Linear polyethylene (LPE) is an important polymer whose relaxation behavior has been subject to differing interpretations (1). One of these centers about the fact that the γ process is skewed toward low temperature and often shows a low temperature shoulder. This has sometimes been interpreted as evidence for the presence of poorly resolved, overlapping competitive processes. The lower temperature tail or shoulder has sometimes been assigned as originating in the crystal phase and the main peak in the amorphous fraction. However phenomenological analysis (2) shows very clearly from relaxation strength variation with crystallinity that the process disappears in the 100% crystalline material and is wholly amorphous phase in origin. The low temperature tail is a simple and direct result of the α width parameter being temperature sensitive. The peak narrows in the frequency domain as temperature increases. If the relaxation time distribution were temperature independent, and relaxation times had Arrhenius temperature dependence the peak would be symmetrical versus 1/T and hence skewed toward high temperature versus T. However the sharpening of the width with increasinfg temperature can easily be sufficient to reverse this effect and lead to a low temperature tail. Most low temperature, sub-glass, " γ"-like processes show this behavior even those in completely amorphous materials. If the broadening of the peak as temperature decreases halts at low temperature and the width becomes frozen-in (as in Figure 3) a shoulder on the peak can be produced. Thus, with the modification of the α parameter temperature behavior incorporated in equation 6 these shoulders are easlily reproduced for linear polyethylene (2).

Acknowledgment

The author is grateful to the Polymers Program, Division of Materials Research, National Science Foundation for financial support of this work.

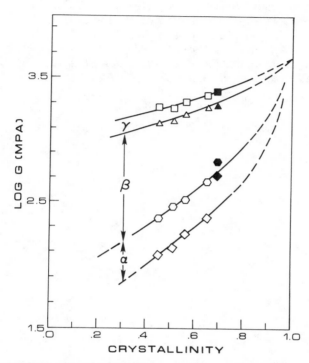

Figure 8. Relaxation strength versus crystallinity in isotactic polypropylenes of Figure 7. Unrelaxed low temperature modulus (□), relaxed γ modulus (△), relaxed β modulus (○), relaxed α modulus (◇). Filled symbols are for the isothermally crystallized (68%) specimen.

## Literature Cited

1. Boyd, R. H. *Polymer*. 1985, 26, 323.
2. Boyd, R. H. *Macromol.*. 1984,17, 903.
3. Mc Crum, N. G.; Read, B. E.; Williams, B. E. "Anelastic and Dielectric Effects in Polymeric Solids"; Wiley: New York,1967.
4. Cole, K. S.; Cole, R. H. *J. Chem. Phys.* 1941,9, 341.
5. Havriliak, S.; Negami, S. *Polymer* 1967, 8, 161.
6. Boyd, R. H.; Aylwin, P. A. *Polymer* 1984,25,340.
7. Passaglia, E.; Martin, G. M. *J. Res. Nat. Bur. Stds.* 1964, 68, 519.
8. Gray, R. W.; McCrum, N. G. *J. Polym. Sci. Part A-2* 1969,8, 1329.

RECEIVED November 14, 1985

# Automated Rheology Laboratory
Part I

**V. G. Constien, E. L. Fellin, M. T. King, and G. G. Graves**

**Dowell Schlumberger, Tulsa, OK 74101**

>A computer-controlled rheology laboratory has been constructed to study and optimize fluids used in hydraulic fracturing applications. Instruments consist of both pressurized capillary viscometers and concentric cylinder rotational viscometers. Computer control, data acquisition and analysis are accomplished by two Hewlett Packard 1000 computers. Custom software provides menu-driven programs for instrument control, data retrieval and data analysis.

Hydraulic fracturing is a method of stimulating production of oil or gas from rock formations. A fluid is pumped under conditions of high pressure and high rate into the formation to fracture it. The fluid also carries sand or a similar proppant material into the fractures. When the pumping is stopped and the hydraulic pressure is released at the wellhead, the fracture partially closes on the sand leaving a highly permeable channel for the oil or gas to flow back to the well.

The fluids which are used in the hydraulic fracturing process can be quite complex. Laboratory research and development on these fluids require many hours of rheology testing to discover suitable compositions and systems. To speed this process, an automated stimulation fluid rheology laboratory was constructed. In this paper, we describe the types of instruments which are used in this laboratory, the computer system, and how typical experiments are set up, run, and results stored and analyzed. In the accompanying paper (Part II), the focus is on the automation of one of the more complex instruments in this laboratory.

**Fluid Compositions**

Typical hydraulic fracturing fluid compositions have been described by several authors (1). The fluids may be aqueous or hydrocarbon base and may also contain energizing gases such as nitrogen or

0097-6156/86/0313-0105$06.00/0
© 1986 American Chemical Society

carbon dioxide. A typical, high-performance, aqueous fluid could have the following components.

- Fresh Water
- Bactericide
- Antifoam
- Potassium Chloride
- Hydroxypropylguar
- Buffer
- Crosslinker
- Fluid-Loss Control Additive
- Polymer Degrading Additive
- Surfactant

The purpose of the various ingredients is to provide desirable fluid properties at different times in the treatment. For example, the crosslinker system may be designed to have a low reaction rate while the fluid is at low temperatures (such as those found in the mixing tanks and before the fluid has traveled very far into the fracture). Then as the fluid temperature increases above a threshold value, the crosslinking rate accelerates and changes the rheology of the fluid to one more desirable in the low-shear, but high-temperature, environment of the fracture.

Unfortunately, from a research time standpoint, slight changes in fluid composition or in treatment conditions (such as pumping rate) can frequently cause significant changes in fluid rheology. To adequately study and optimize fluid rheology often requires many hours of composition vs performance testing at several different conditions.

## Laboratory Instruments and Computers

The configuration of the laboratory is shown in Figure 1. The rheology equipment consists of different types of viscometers which are used by research and development personnel to measure properties of hydraulic fracturing fluids. Some of these instruments are not commercially available and were built "in-house" to simulate specific treatment conditions.

To subject test fluids to the full range of shear, temperature and time conditions which may be encountered in a typical fracturing treatment, different instruments may be coupled together. For example, a capillary viscometer with in-line additive pumps is used to continuously prepare fluids and to simulate the shear history a fluid would experience as it travels down the tubular goods on its way to the formation to be fractured. Once the fluid enters the formation, different shear, temperature and time conditions are encountered. These conditions are simulated in the laboratory by loading a portion of the fluid (exiting the shear history simulator) directly into a high-temperature/high-pressure rotational viscometer or a reciprocating capillary viscometer for long-term rheology evaluations (Figure 2) (2).

The computer system for this laboratory consists of a single Hewlett Packard 1000 A-Series minicomputer which is networked to an HP 1000 F-Series host computer. The A-Series computer was chosen because its operating system (RTE-A) is real-time, multiuser and

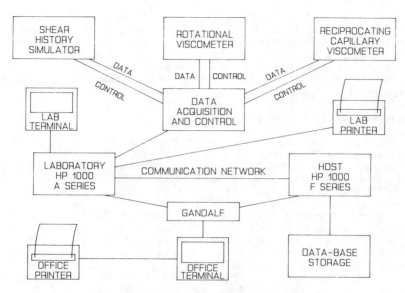

Figure 1. Instrument and computer network.

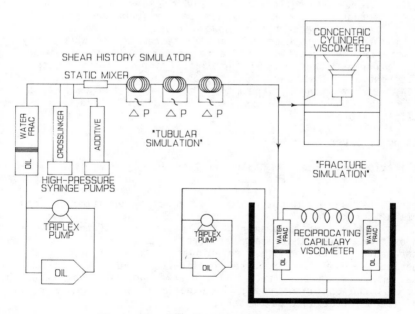

Figure 2. Rheology instruments.

multitasking. This operating system allows multiple users and multiple programs to be run at the same time, with data collection and control programs having the highest priority. Upon completion of the experiment, initial data analysis is done on the A-Series computer. Data may then be transferred into project data bases on the HP 1000 F-Series host computer where more powerful data analysis software is available.

## Computer Interface with Instruments

Inputs (and outputs) from the instruments to the computer are handled by two HP 3497A data acquisition and control units. Each HP 3497A has a combination of the following input/output cards based on the instrument interfacing requirements: 44421A 20-Channel Guarded Acquisition, 44425A 16-Channel Isolated Digital Input, 44428A 16-Channel Actuator Output and 44429A Dual Voltage D/A. Each HP 3497 is interfaced over a dedicated Hewlett Packard Interface Bus (HPIB) to an HP 1000 A-Series minicomputer with 1.5 megabytes of memory and 27 megabytes of permanent disc storage. Each HP 3497 is operated by a set of four programs. The first program runs at bootup and performs initialization functions. The second program is scheduled to run every second. Its purpose is to determine if any instruments require data collection (from the status of flags set in system common). If so, it schedules the third program as the fastest rate required for data collection on any one instrument. The third program "programs" the HP 3497 for analog input and schedules the fourth program to accept the input voltages from the HP 3497. In addition to placing these voltages in system common where they can be accessed by specific application programs, the fourth program also "programs" the HP 3497 for digital input, digital output and analog output. Instrument application programs communicate with the HP 3497 programs through the memory resident system common. System common is also used to pass real-time control parameters between application programs.

## Setting Up Experiments

Application programs are custom written in FORTRAN for each set of "like" instruments. For ease of use, the programs are accessed through a main program menu. There is one menu for each set of instruments. Program options are selected by pressing the desired terminal softkey.
   To set up an experiment, the researcher defines the fluid composition and instrument control parameters. For some instruments, such as the rotational coaxial viscometers, the experiment setup can be quite complicated. For this reason, all data entry is of the "fill-in-the-blank" nature. The researcher also has the option of using the experiment setup from a previous experiment for default parameters.

## Computer Control and Data Acquisition During Experiments

Once the experiment setup information has been successfully entered, the researcher can choose a menu option to start data collection and/or control on the instrument of his choice. The data collection and control programs run in the background leaving the terminal free for other menu selections. At this time, the researcher is free to set up another experiment, analyze the data from a previous experiment, etc.

Instruments are controlled by information contained in the experimental setup file. For each type of instrument (shear history simulator, rotational viscometer, reciprocating capillary viscometer), the hardware is controlled so that the parameters of shear rate, temperature and time comply with the desired test conditions. This involves controlling devices such as pumps, bath heaters, valves and variable-speed motors. The setup and control parameters are recorded in the experiment file along with the resulting measured data. If necessary, the experiment can easily be repeated.

While the experiment is running, informational messages are logged to a printer designated for that purpose. Real-time data (temperatures, pressures, etc.) can be displayed using laboratory or office terminals. The researcher can also view the data analysis results for the latest set.

## Data Analysis Methods

Hydraulic fracturing fluids are solutions of high-molecular-weight polymers whose rheological behavior is non-Newtonian. To describe the flow behavior of these fluids, it is customary to characterize the fluid by the Power Law parameters of Consistency Index (K) and Behavior Index (n). These parameters are obtained experimentally by subjecting the fluid to a series of different shear rates ($\dot{\gamma}$) and measuring the resultant shear stresses ($\tau$). The slope and intercept of a log shear rate vs log shear stress plot yield the Behavior Index (n) and Consistency Index ($K_v$), respectively. Consistency Indices are corrected for the coaxial cylinder viscometers by:

$$K = K_v * \left( \frac{\beta^{2/n}(\beta^2 - 1)}{n(\beta^{2/n} - 1)\beta} \right)^{-n}$$

where $\beta$ is the ratio of the inside cup radius to the bob radius.

Consistency Indices are calculated for pipe ($K_p'$) and slot ($K_s'$) geometries (3) in English units of lb-sec$^n$/ft$^2$ and viscosities (n, cp) calculated for 37.7, 170 and 511 sec$^{-1}$ shear rates as follows.

$$K_p' = K \left( \frac{3n + 1}{4n} \right)^n$$

$$K_s' = K \left( \frac{2n + 1}{3n} \right)^n$$

$$n(cp) = \frac{100 * K_p' \dot{\gamma}^{(n-1)}}{478.8}$$

Consistency Indices for the reciprocating capillary viscometers are calculated in a similar manner except that $K_p$ is determined directly from log shear rate vs log shear stress data.

The shear history simulators operate at a single shear rate during an experiment and do not run shear ramps. For these instruments, apparent viscosity at a single shear rate is determined by the relationship of differential pressure ($\Delta P$), capillary length (L) and radius (r), and volumetric flow rates (Q), as follows.

$$\eta = \frac{\frac{\Delta P r}{2L}}{\frac{4Q}{\pi r^3}} = \frac{\text{Shear Stress}}{\text{Shear Rate}}$$

Additional information concerning automation and operation of this instrument is given by King (<u>4</u>) et al. in an accompanying paper.

## Data Storage and Retrieval

The data from each experiment are stored in individual files on the local HP 1000 computer. These files are formatted so that they can be transferred over the laboratory network (using the Hewlett Packard DS/1000-IV networking software) and loaded directly into IMAGE/1000 relational data bases organized by project. Data can then be retrieved, cross-referenced and reported using custom-written search programs (such as rheology vs fluid composition or test conditons) or the data-base query language ASK (from Corporate Computer Systems). The plotting program Grafit (Grafic User Systems, Inc.) is linked directly to the custom search programs, so that figures can be generated automatically from the retrieved data.

## Example Experiment

The following are results from an actual experiment in which experimental conditions, data collection and data reporting were conducted by computer.

### Simulated Hydraulic Fracturing Conditions

| | |
|---|---|
| Well Depth | 10,000 ft |
| Pump Rate | 12 BPM |
| Tubing Size | 2.441 in. ID |
| Tubing Shear Rate | 1,350 sec$^{-1}$ |
| Time at Tubing Shear | 5.0 min |
| Fracture Shear Rate | 170 sec$^{-1}$ |
| Time at Fracture Shear | 227 min |
| Fluid Temperature in Fracture | 300°F |
| Fluid Shut-in Time before Flowback | 12.5 hr |

In this experiment, a Tubing Shear History Simulator was coupled with a Reciprocating Capillary Viscometer to simulate the above conditions. Results from the experiment are given in Tables I and II and Figure 3, and were retrieved directly from the project data base. Total instrument use time for this experiment was 17 hr, of which 16.5 hr were completely unattended operation. Data analysis, including plotting of figures, required less than five minutes.

## Conclusions

Automation and data collection from complex laboratory equipment have been accomplished. The result of this effort has been more efficient use of the researcher's time, improved data analysis and the capability to easily conduct lengthy experiments without personnel being present.

### Table I. Shear History Data

Experiment Identification: 3613-8-65    Date: 10-20-83
Instrument: MISS MO
Comments: Shear History Simulation
Final Temperature (°F): 75
Fluid Type: EXPER.
Composite (polymer) Lot No.: 3401-38-A
Master Batch No.: 3613-4-5

#### Additives in Master Batch

| Additive | Conc. | Units |
|---|---|---|
| ADD1 | 55.000 | lb/1,000 gal |
| ADD3 | .250 | gal/1,000 gal |
| ADD4 | 10.000 | lb/1,000 gal |

#### Additives Added In-Line

| Additive | Conc. | Units |
|---|---|---|
| ADD2 | 83.000 | lb/1,000 gal |

| | Temp. (°F) | pH | Shear Rate | Apparent Viscosities | | |
| | | | | Coil 1 | Coil 2 | Coil 3 |
|---|---|---|---|---|---|---|
| 1 | 75 | 4.8 | 1,348 | 22 | 20 | 20 |

Table II. Long-Term Rheology Data

Experiment Identification: 3613-8-64  Date: 10-20-83
Instrument: FERN
Comments: Continuous shear 4 hr; shut-in 12 hr at temperature
Fluid Type: EXPER.
Composite (polymer) Lot No.: 3401-38-A
Master Batch No.: 3613-4-5
Final Temperature (°F): 300
pH at Room Temperature: 4.8
Shear Rate ($sec^{-1}$): 1,350
Shear Time (min): 5.0

| Additive | Conc. | Units |
|---|---|---|
| ADD1 | 55.000 | lb/1,000 gal |
| ADD2 | 83.000 | lb/1,000 gal |
| ADD3 | .250 | gal/1,000 gal |
| ADD4 | 10.000 | lb/1,000 gal |

| Time (min) | Temp. (°F) | n' | Kp | CORR | Ks | Viscosities | | |
|---|---|---|---|---|---|---|---|---|
| | | | | | | 37.7 | 170 | 511 |
| -33.0 | 81 | 0.1742 | 0.2039 | 0.9658 | 0.2099 | 487 | 141 | 57 |
| 0.0 | 282 | 0.1790 | 0.0598 | 0.9503 | 0.0616 | 145 | 42 | 17 |
| 30.0 | 299 | 0.1994 | 0.0520 | 0.9879 | 0.0536 | 136 | 41 | 17 |
| 58.0 | 300 | 0.1845 | 0.0683 | 0.9178 | 0.0703 | 169 | 50 | 20 |
| 114.0 | 300 | 0.1364 | 0.1260 | 0.9940 | 0.1292 | 263 | 71 | 28 |
| 180.0 | 300 | 0.1235 | 0.1373 | 0.9178 | 0.1406 | 273 | 73 | 28 |
| 227.0 | 300 | 0.1405 | 0.1190 | 0.9946 | 0.1221 | 252 | 69 | 27 |
| 975.0 | 300 | 0.3038 | 0.0222 | 0.9846 | 0.0229 | 85 | 30 | 14 |

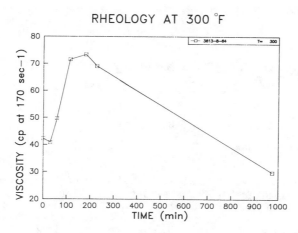

Figure 3. Experimental results.

**Literature Cited**

1. M. W. Conway, S. W. Almond, J. E. Briscoe and L. E. Harris: **J. Pet. Tech.** (Feb. 1983), 315.
2. L. J. Craigie: "A New Method for Determining the Rheology of Crosslinked Fracturing Fluids using Shear History Simulation," SPE/DOE 11635 presented at the 1983 SPE/DOE Symposium on Low Permeability (March 14-16), Denver, Colo.
3. J. G. Savins: "Generalized Newtonian (Pseudoplastic) Flow in Stationary Pipe and Annuli," **Pet. Trans.**, AIME (1958), **213**, 325.
4. M. T. King, V. G. Constien and E. L. Fellin: "Automated Rheology Laboratory--Part II: An Automated Instrument for Continuous Rheology Evaluation," submitted to the ACS Symposium, Division of Polymeric Materials: Science and Engineering, Computers in Applied Polymer Science: IV (1985), Miami Beach.

RECEIVED March 4, 1986

# 11

## Automated Rheology Laboratory
## Part II

**M. T. King, V. G. Constien, and E. L. Fellin**

**Dowell Schlumberger, Tulsa, OK 74101**

As a component of the automated rheology laboratory described in Part I, an instrument which is used in a method to evaluate water-soluble polymer systems has been constructed, and then automated by being interfaced to a minicomputer. With this instrument, which measures apparent viscosities, fluid composition can be easily changed by sequencing through different types and concentrations of additives and polymers. This permits the rapid screening and optimization of a large number of additive and polymer candidates under varying conditions. The instrument is capable of continuous evaluation of any number of fluids and additives at eight temperatures from 24° to 149°C and at shear rates from 170 to 1,350 sec$^{-1}$. Automation is accomplished through the use of a Hewlett Packard 1000 and includes instrument operation, data collection, reporting and archiving results.

Research in search of new and improved hydraulic fracturing fluids is important in companies which provide fracturing treatments for oil and gas wells. The fracturing fluids typically consist of a natural or synthetic polymer thickening agent dissolved in a light brine, together with additives such as crosslinkers, stabilizers and/or degrading agents which combine to produce the desired properties for a specific set of well conditions. Considerable time is spent in the selection of thickening agents and the optimization of fluid systems to produce the desired rheology. One of the reasons that so much time is spent is the method used. The purpose of this research was to develop a less time-consuming method which also yields more repeatable data.

The conventional method and instrumentation for hydraulic fracturing fluid evaluation have evolved over several years. This method is usually based upon traditional experimental design where each variable is changed in turn while the others are held constant. Although a vast amount of rheological data on hydraulic fracturing fluids has been generated this way, the conventional method is very time consuming for several reasons. First, a large number of samples must be individually prepared (usually in a one-quart, kitchen-type blender). Then they must be loaded on a coaxial cylindrical rheometer for long-term rheology measurements (usually one to eight hours). Due to poor performance, many of these samples will not be tested for the full time; however, a large amount of time is spent cooling down the sample and setting up the instrument for the next test. In addition, if the rheometers do not have some means of automatic data collection, still more time is spent analyzing and storing the data. Although some time savings can be realized by using some type of experimental design, the time-consuming factors are still the sample preparation and the type of rheometer used.

To reduce the time needed for fluid evaluation, a new testing method and an associated instrument have been developed. This method saves time by utilizing an automated, continuous, variable temperature, single-pass, pipe rheometer capable of adding additives in-line. This eliminates both the need for the large number of individually prepared samples and a large part of the time spent using the conventional cylindrical rheometer.

## Approach

Basically, the method consists of preparing a large quantity of a solution of a polymer dissolved in water. As this solution is pumped through the rheometer, additives are added and mixed in-line. The rheology of the resulting fluid is then determined by measuring the differential pressure across a series of coiled capillary tubing, while the fluid is being pumped under preset conditions of shear and temperature. Fluid compositions can be easily changed by sequencing through different additives and different amounts of these additives. This method can be used to rapidly study the effects of additives or conditions upon a particular polymer system or can screen a large number of polymers or additives looking for promising candidates. Comparisons of different fluids are usually made based on their apparent viscosities calculated using the following equations:

$$\dot{\gamma} = \text{Shear Rate in sec}^{-1} = \frac{32\ Q}{\pi D^3}$$

$$\tau = \text{Shear Stress in dynes/cm}^2 = \frac{D\Delta P}{4L}, \text{ and}$$

$$\eta = \text{Apparent Viscosity in poise} = \frac{\text{Shear Stress}}{\text{Shear Rate}} = \frac{\tau}{\dot{\gamma}}$$

where  Q = Flow Rate, mL/sec,
       D = Pipe Diameter, cm,
       ΔP = Frictional Pressure Drop, dynes/cm$^2$, and
       L = Pipe Length, cm.

**Instrument Design**

Figure 1 illustrates the basic design of the instrument. One of three test fluids is loaded into the upper portions of two floating piston accumulators. Hydraulic fluid is pumped at a selected flow rate into the bottom of one of the accumulators forcing the piston upward and displacing the test fluid into the tubing. The flow rate can be selected to give a shear rate in the range of 170 to 1,350 sec$^{-1}$ (10 to 80 mL/min through 0.22-cm diameter tubing). Additives are added in-line using high-pressure metering pumps. Up to three additives may each be added at rates of 0.8 to 200 mL/hr. Fluid pH is monitored with an in-line pH probe mounted directly in the flow line. The fluid can be given a shear history by passing it through one or more of the tubing coils at the specified pump rate. Different shear histories possible range from a maximum of 36 min at 170 sec$^{-1}$ to a minimum of 1.5 min at 1,350 sec$^{-1}$. The fluid can then be loaded into other rheometers for long-term rheology testing; or if preliminary screening is being conducted, the fluid can be sent to a heated coil which is located in a variable temperature bath. Comparisons of the resulting apparent viscosities allow the researcher to determine which fluid compositions appear to be candidates for additional long-term rheology testing.

**Computer System**

Automation is accomplished by the sequential control of various electrical devices using an HP 1000 computer system. The HP 1000 is part of a distributed network of HP 1000s dedicated to laboratory automation at Dowell Schlumberger. Figure 2 contains the portion of the network associated with this instrument. There are two HP 1000s concerned directly with data acquisition, storage and retrieval. The HP 1000 located in the laboratory is a minicomputer with 1.5 megabytes of memory and 27 megabytes of permanent disc storage. It provides program control of an HP 3497A Data Acquisition and Control Unit which has five card slots containing two 16-channel actuator output cards, one 16-channel isolated digital input card, one 20-channel relay multiplexer and one dual voltage output (0 to ± 10V) D/A converter. The local HP 1000 is linked through Hewlett Packard's distributed networking software (DS/1000-IV) to an HP 1000 F-series host computer located in a computer room some distance from the laboratory.

The host HP 1000 is a minicomputer with two megabytes of memory and 644 megabytes of permanent disc storage. This computer provides a means for data storage and retrieval. The data are stored in IMAGE/1000 data bases and can be retrieved and reported using either ASK (Corporate Computer Systems) routines or application-specific search programs. Researchers can gain access to either of the HP 1000s through terminals located conveniently in both the laboratory and offices. An HP 2382A terminal connected directly to the laboratory computer and located near the instrument is used primarily to

11. KING ET AL.  *Automated Rheology Laboratory: Part II*

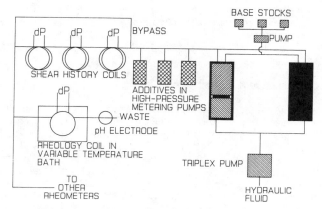

Figure 1. Basic design of instrument.

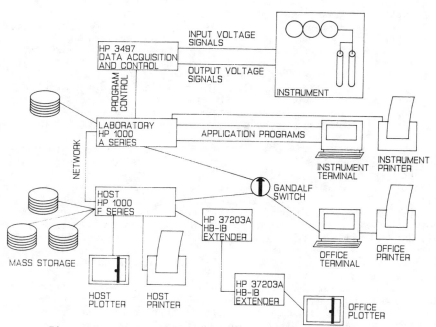

Figure 2. Instrument and associated computer network.

gain access to the instrument control programs. An HP 2671A printer, also connected directly to the laboratory computer, provides error and data reporting. Office terminals, such as the HP 150, can access either computer through a Gandalf PACX4 data switch. An HP 9872 plotter is used for graphical displays and overhead slide preparation. An HP 2608A on the host HP 1000 is used for large data-base printouts.

## Computer and Instrument Interface

Four different types of tasks are performed by automation. Two involve the sequencing of valves and pumps involved in the setup and completion of the designed experiment through the operation of the test and hydraulic fluid systems. The other tasks involve the control of the temperature bath and data collection. To perform these tasks, air-actuated solenoids and optically coupled solid-state relays are used. These devices are controlled by an electrical circuit consisting of the device connected in series with a power supply and a channel on the actuator card in the HP 3497. The power supply is either 24 VDC for use with the solenoids or 5 VDC for the solid-state relays. The actuator output channel acts as a simple on/off switch which allows power to be supplied to the solenoid or relay when closed. The logic of the circuit is controlled by application programs running on the local HP 1000.

The direction and flow rate of the test and hydraulic fluids are determined by nine three-way valves and six air-driven hydraulic pumps that must be sequenced in the proper order. The position of the valves is determined by six air-driven actuators. Two of the pumps are miniaturized, air-driven, hydraulic pumps used for sample loading and pressurization. One of the remaining four pumps is a high-pressure, constant volume, positive displacement, piston metering pump to provide hydraulic pressure, and the other three are positive displacement syringe pumps for in-line addition of additives.

The temperature of the bath is controlled by a built-in heater or refrigerated coolant to maintain the selected temperature. Because it was not possible to externally set the bath's built-in controller with our computer, it was necessary to make our own circuit. This circuit consists of an analog multiplexer, eight reed relays and eight multiturn potentiometers. This external resistance network along with the bath's internal components forms a voltage divider network to generate the temperature set point. The circuit determines which resistance to apply from a 5 VDC, three-bit, binary code supplied by the HP 3497. The three-bit binary code allows eight temperatures to be selected.

Pressure measurements are made with four differential pressure transducers which produce a 0- to 5-VDC output signal from an unregulated 24-VDC input. The range of three of the transducers is 0 to 500 psi and the other is 0 to 320 psi. The signal from these transducers is connected to an analog input card in the HP 3497.

## Application Programs

A set of FORTRAN programs written specifically for the instrument is available to aid in setting up an experiment, running the instrument, collecting data during the experiment, analyzing and reporting the results and, finally, archiving the data in a data base for future data analysis and reporting. All of the programs are accessed through a main menu program which is available on any terminal connected to the local HP 1000 computer system.

When the main menu program is requested, the first page of the two-page menu will be displayed on the local terminal screen (see Figure 3). At this point, any of the eight menu options can be selected. The selection is done using the programmable softkeys on the terminal, making the menu user-friendly and easy to operate.

Whenever a test is to be run, the sample composition and instrument control parameters must be defined. This is done with three (or more) data-entry screens. The first data-entry screen, shown in Figure 4, deals with experiment identification and base fluid composition. The operator simply types in the desired information into unprotected fields of the screen. Information requested includes such items as experiment ID, submitter's name, base fluid type and base fluid additives. The base fluid pump rate and valve selection are also requested for later use by the control programs. The second data-entry screen is used to select the desired test temperatures and also to record any comments related to the experiment. The third data-entry screen is used to input the in-line additive compositions. This screen is filled out for each set of additives to be tested with the base fluid as described on Data-Entry Screen No. 1. Also input are the pump rates for each of the three additive pumps. This information is used by the control programs when the additive set is being tested. (The pump rates are preset by the operator, but the pumps are turned on and off by the control programs as necessary during the course of an experiment.)

Once the experiment setup information has been successfully entered, it is stored in an ASCII file. This experiment setup file can then be placed in the queue of experiments to be run on this instrument or it can be saved for later use with the manual data collection program. Any of the information entered in the experiment setup phase can be changed with the experiment setup editor. The editor is accessed through the main menu program and works the same as the initial experiment setup program.

The screening capillary viscometer can be operated manually as well as automatically. If manual operation is desired, an interactive program is available to aid with data collection. Program-operator interaction takes place through terminal input and also with a push-button data collection indicator on the instrument itself. The immediate on-line analysis of results and the ease of data storage and retrieval are just some of the benefits realized by using this program in conjunction with manual operation of the instrument.

There are three programs involved in the automatic operation of the instrument. The main control program handles the sequencing and timing operations necessary to run each experiment. It also collects the data when required. A second program is responsible for switching and filling the accumulators to ensure a continuous flow of base fluid. The last program controls the bath temperature.

During an experiment, the temperature and pressure are monitored for any over-temperature or over-pressure conditions which might occur. If an error condition were to occur, the instrument

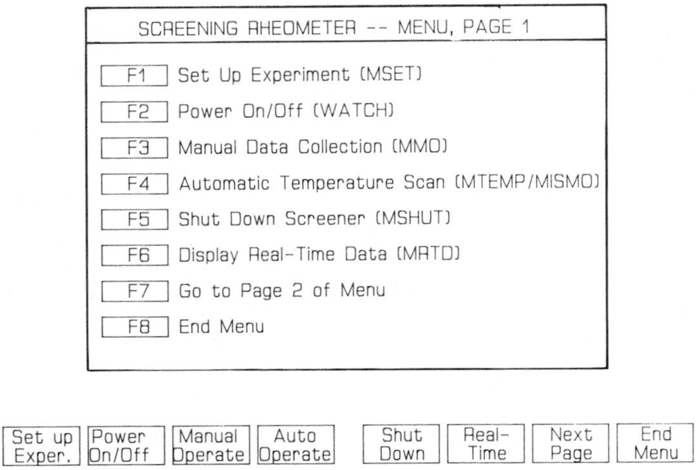

```
┌─────────────────────────────────────────────────┐
│       SCREENING RHEOMETER -- MENU, PAGE 1       │
│                                                 │
│     [ F1 ]  Set Up Experiment (MSET)            │
│     [ F2 ]  Power On/Off (WATCH)                │
│     [ F3 ]  Manual Data Collection (MMO)        │
│     [ F4 ]  Automatic Temperature Scan (MTEMP/MISMO) │
│     [ F5 ]  Shut Down Screener (MSHUT)          │
│     [ F6 ]  Display Real-Time Data (MRTD)       │
│     [ F7 ]  Go to Page 2 of Menu                │
│     [ F8 ]  End Menu                            │
│                                                 │
└─────────────────────────────────────────────────┘
```

| Set up Exper. | Power On/Off | Manual Operate | Auto Operate | Shut Down | Real-Time | Next Page | End Menu |

Figure 3.  One of the menu screens.

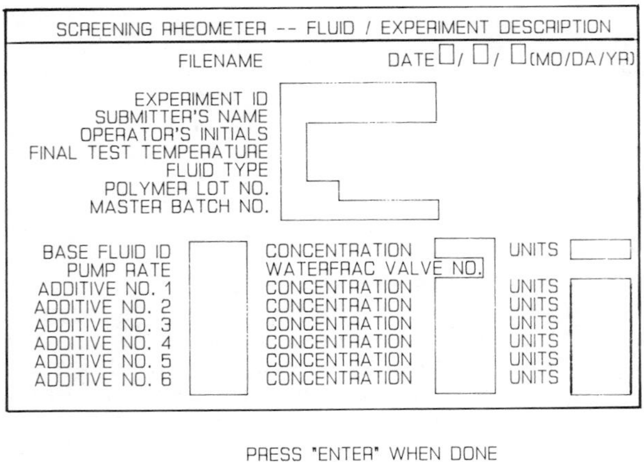

```
┌──────────────────────────────────────────────────────┐
│   SCREENING RHEOMETER -- FLUID / EXPERIMENT DESCRIPTION │
│                FILENAME         DATE □/ □/ □(MO/DA/YR)  │
│         EXPERIMENT ID    ┌──────────┐                │
│       SUBMITTER'S NAME   │          │                │
│      OPERATOR'S INITIALS │          │                │
│   FINAL TEST TEMPERATURE │          │                │
│              FLUID TYPE  │          │                │
│          POLYMER LOT NO. │          │                │
│         MASTER BATCH NO. └──────────┘                │
│                                                      │
│   BASE FLUID ID □   CONCENTRATION □      UNITS □    │
│       PUMP RATE     WATERFRAC VALVE NO.              │
│    ADDITIVE NO. 1   CONCENTRATION        UNITS      │
│    ADDITIVE NO. 2   CONCENTRATION        UNITS      │
│    ADDITIVE NO. 3   CONCENTRATION        UNITS      │
│    ADDITIVE NO. 4   CONCENTRATION        UNITS      │
│    ADDITIVE NO. 5   CONCENTRATION        UNITS      │
│    ADDITIVE NO. 6   CONCENTRATION        UNITS      │
└──────────────────────────────────────────────────────┘
```

PRESS "ENTER" WHEN DONE

Figure 4.  Experiment description screen.

would be shut down in a safe and orderly manner, then the power to the instrument would be turned off. In addition, a message would be printed on the error logging printer designated for this purpose.

It is sometimes necessary to shut down control on the instrument before it finishes its queue of experiments. The shutdown program can be activated from the main program menu. Once activated, the instrument is shut down in a safe and orderly manner to be sure everything gets reset and/or turned off and the control programs terminated.

Once a test is complete, another menu option provides the data analysis results in the form of a hard-copy report printed on the local line printer. The report includes experiment identification information and the apparent viscosities calculated for each data set. A subset of the data analysis program is scheduled automatically by the control programs while the experiment is in progress and provides immediate on-line analysis of apparent viscosities for each data set as it is collected. The results are viewed using the real-time data display program (Figure 5).

At any time during the experiment, the researcher can view a real-time display of the instrument's data. These data include the current sample temperature, the current sample pH and the current delta pressure readings. Also displayed is the status of all digital inputs (pumps, valves, etc.), the data analysis results from the latest data set and the experiments in the queue waiting to be run on the instrument. These real-time data are updated approximately once per second with the entire display being refreshed approximately every 30 seconds.

Since the data base for this instrument resides on the host HP 1000 computer, the experiment setup files must first be transferred from the local computer to the HOST computer. This is done using the Dowell Schlumberger local laboratory computer network and the Hewlett Packard DS/1000-IV networking software. The programmatic user interface to the network is again accessed through the main menu program for the instrument.

## Example Data

In Figure 6, apparent viscosity is plotted as a function of temperature for a typical experiment. The purpose at this experiment was to identify the change in apparent viscosity caused by the addition of an additive at three different levels and eight temperatures. The time required for this experiment using the automated instrument was about four hours. The same experiment would have required about 80 hours using the conventional method.

## Conclusions

Development and use of this new testing method and instrument for evaluating thickening agents based upon an automated pipe rheometer have proved to significantly reduce the time required for these evaluations.

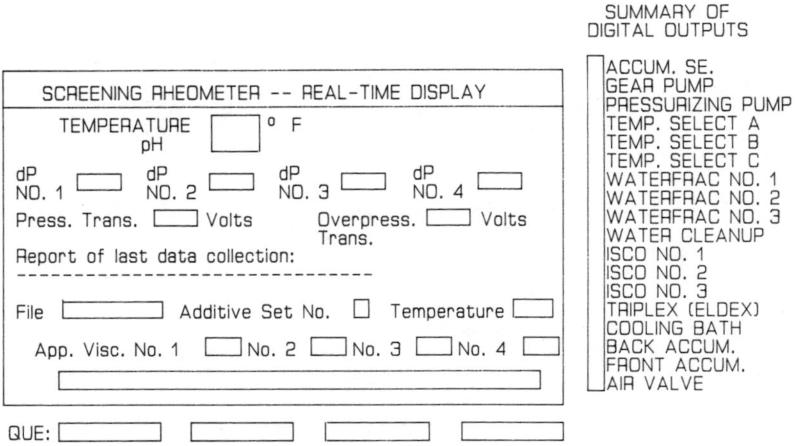

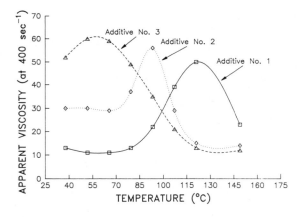

Figure 5. Real-time display screen.

Figure 6. Data from a typical experiment.

RECEIVED March 4, 1986

# 12

# An Automated Analysis System for a Tensile Tester

T. T. Gill and Mark E. Koehler

Glidden Coatings and Resins, SCM Corporation, Strongsville, OH 44136

> An Instron Tensile Tester Model TM was interfaced to a micro-computer for data collection and transmission to a minicomputer. A FORTRAN program was developed to allow data analysis by the minicomputer. The program generates stress-strain curves from the raw data, calculates physical parameters, and produces reports and plots.
>
> The Instron automation provides an easy and rapid means for acquiring accurate analyses of tensile properties. Sample preparation is now the rate limiting step as opposed to data analysis.

Tensile testing is an important part of the physical characterization of free film coatings. The fundamental properties measured relate directly to performance properties of the coating. Because of the time required to obtain and analyze tensile data, a laboratory which routinely performs tensile tests may find that an automated system is needed. Although commercial packages are available, it is feasible to develop an in-house system with relatively little expense. This paper describes one such system as implemented at Glidden Coatings and Resins with very satisfactory results.

System Configuration

A dedicated microcomputer is interfaced to the Instron instrument in order to collect the raw data. The microcomputer consists of an 8080A microprocessor, 32K bytes of memory, A/D converter, serial I/O for communication, parallel I/O for digital control and sensing, a real-time programmable clock, and vectored interrupt control.

The microcomputer is first initialized by means of a DIALOG program on the minicomputer which transmits calibration information, sample identification, and data collection rates. The microcomputer collects data at two operator selectable rates, initial and final. The starting rate is set faster to provide better initial resolution, with typical initial and final rates of 100 pts/sec and 10 pts/sec, respectively. The duration of the initial period can also be varied, with a usual length of 1000 points.

Once the microcomputer has received this information, a ready light is turned on and data acquisition can be begun by pushing a start button. A busy light then flashes to indicate that data collection is in progress. After the sample breaks, the operator signals the end of data collection by pushing a stop button. The microcomputer then transmits the data to the minicomputer and re-initializes for another run. Meanwhile, the minicomputer stores the data in a unique job file. Details regarding the minicomputer system have been previously reported (1).

Data Analysis Program

Instron data files which have been saved on the minicomputer can be analyzed at any time using the STRESS program. After reading the specified raw data file, the STRESS program first converts the voltages to stress values based upon the data collection rates and calibration parameters, according to Equation 1.

$$S(I) = \frac{F(V(I)-Co)}{WT(Cl-Ch)} \qquad (1)$$

In Equation 1, S is the stress at point I in PSI, V is the raw voltage data for point I in millivolts, F is the full scale load in pounds, W is the sample width and T is the thickness in inches, and Co, Cl, and Ch are calibration data for zero, load, and sample hanger in millivolts.

The program then locates the starting point of the stress-strain curve. The most dependable method for identification of the starting point will vary with sample and instrument behavior. For our coatings work, we search for a datum greater than zero followed by two successively higher values. Variants of this approach have yielded inaccurate starting points which cause considerable error in the Young's Modulus (initial slope) computation.

Next, strain values are calculated according to the following equations. Equation 2 is used in the initial collection rate period and Equation 3 is used in the final period.

$$Ei(I) = \frac{CTi(I-Is)}{60000L} \qquad (2)$$

$$Ef(I) = \frac{C(Tf(I-Ir)+Ti(Ir-Is))}{60000L} \qquad (3)$$

In Equations 2 and 3, Ei and Ef are strains at point I in the initial and final collection rate areas, Is is the starting point and Ir is the rate transition point, Ti and Tf are the initial and final collection rates in sec/point, C is the crosshead speed in inches/sec, and L is the sample length in inches.

After these calculations, a least squares fit is done on the first 250 milliseconds of data to obtain the initial slope, which is Young's Modulus. This represents 5% or less of the data for a run of at least 2.4 seconds at an initial data collection rate of 100 points per second.

STRESS searches for the Yield Strength (maximum stress), then continues on to find the greatest drop in stress. From there it reverses direction to look for a peak or a point 0.004 L/Lo prior, whichever comes first. This point is taken to be the Break Point.

The value of 0.004 L/Lo was chosen empirically since most breaks do not exceed this range.

After the curve is characterized, the Work at Break (Wb) is calculated as the integral of stress (S) as a function of strain (E) from the start point (1) to to break point (n).

$$Wb = \int_{E1}^{En} S(E)dE \quad (4)$$

If replicate runs were made, STRESS will repeat the preceeding calculations for each data set.

## Fracture Analysis

The Instron analysis program also can perform a Fracture Analysis if multiple runs are made as a function of flaw length. To perform a Fracture Analysis, the operator indicates this in the DIALOG session. The DIALOG will ask for the flaw lengths for each run. The STRESS program recognizes this flag and calculates fracture energy after all runs have been processed. Two relationships are utilized, depending on whether the sample exhibits elastic or viscoelastic behavior. The latter is defined as materials with a strain at break greater than 2%. The following equations show fracture energy to be equal to the slope of a line having an intercept of zero, based upon the work of Griffith (2).

$$Wb = \alpha e \frac{1}{AY^2} + 0 \quad (5)$$

$$\sigma^2 = \alpha v \frac{2E}{AY^2} + 0 \quad (6)$$

$$Y = 1.99 - 0.41(A/W) + 18.7(A/W)^2 - 38.48(A/W)^3 + 53.85(A/W)^4 \quad (7)$$

Where $\sigma$ equals tensile strength, A equals one half flaw length, Y is an edge correction (3), and $\alpha$ equals fracture energy. Slope is computed by least squares analysis with correlation coefficient of fit.

## Outlier Identification

Tensile testing is susceptible to invalid runs due to sample flaws, poor mounting, or many other other sources of error. It is therefore essential that outliers be identified and removed. After all runs have been analyzed, STRESS calculates means and standard deviations for each parameter. It also performs Student t-tests on all results in order to help identify outliers. The Student t-score is calculated using Equation 8.

$$t = \frac{(X_v - V)n^2}{V} \qquad (8)$$

In Equation 8, V is the value to be tested, $X_v$ is the mean, and n is the number of values. The t-score is converted to a probability by a subroutine which calculates the incomplete Beta function. This could also be accomplished using an internal look-up table. Any results with a probability of deviation in the 5-10% range are highlighted with one asterisk. Those with a probability of deviation of 0-5% are highlighted with two asterisks.

A second method used to verify the validity of the results is visual inspection of the graphic data. After the tabular results are presented, the operator can call a subroutine which plots the stress-strain curves. Anomalous curves can usually be easily identified in this manner.

### Editing and Reporting Capability

After inspecting the tabular and graphic data, the operator is allowed to remove runs which appear to be outliers. Any run can be deleted or restored in any order, and the comparative statistics are recalculated with each operation. By comparing the standard deviation before and after deleting a run, the effect of that run can by determined. The editing process can continue indefinitely until the operator is satisfied with the validity of his results.

Once satisfied with the results, the operator can print out a report giving experimental parameters and results of the analysis. A sample report is given in Figure 1.

### Plotting Capability

Any plot which is formed during data analysis can be accumulated and all accumulated plots can be printed out after the analysis is complete. An example of a standard run is shown in Figure 2. For fracture studies, the fracture data can also be plotted along with the least squares line which yielded the Fracture energy.

### Conclusions

The Instron data collection and analysis system described has several strong features. First, it is tailored to our needs and designed to take maximum advantage of existing system utilities, such as the plotting package and the subroutine for obtaining exact probabilities from the t-statistic. It is amendable to alteration or expansion if needs should change or grow. Only the addition of a robotic sample preparation and loading system would be required to completely automate our Instron analyses([4]). Operator interaction is minimized wherever possible, yet the program helps the operator to make judgement calls regarding data validity. The automated system improves the efficiency of the instrument and operator and enhances the aesthetic impression of Instron testing.

Experimental Parameters
Job Number: 20102

| | | | |
|---|---|---|---|
| Sample ID: | 879C602G Control | | |
| Operator ID: | PJM | Run Date: | 16-Apr-85 |
| Initial Period: | 10ms | Final Period: | 100ms |
| Crosshead Speed: | 2.000in/min | Full Scale Load: | 20.lbs |
| Sample Length: | 1.000in | Sample Width: | 0.750in |
| Temperature: | 72.00°F | Humidity: | 50.00% |

| Run | Thickness (in) |
|---|---|
| 1 | 0.003700 |
| 2 | 0.004600 |
| 3 | 0.005000 |
| 4 | 0.003700 |
| 5 | 0.004400 |
| 6 | 0.003600 |

Results of Analysis

| Run Number | Youngs Modulus | Yield Strength (PSI) | Stress At Break (PSI) | Strain At Break (% Lo) | Work At Break (InLb/CuIn) |
|---|---|---|---|---|---|
| 1 | 0.410E+05 | 0.279E+04 | 0.279E+04 | 37.80 | 0.709E+03 |
| 2 | 0.397E+05 | 0.334E+04 | 0.334E+04 | 51.13 | 0.976E+03 |
| 3 | 0.400E+05 | 0.295E+04 | 0.295E+04 | 44.50 | 0.812E+03 |
| 4 | 0.451E+05 | 0.257E+04 | 0.257E+04 | 33.77 | 0.627E+03 |
| 5 | 0.437E+05 | 0.269E+04 | 0.269E+04 | 39.10 | 0.685E+03 |
| 6 | 0.456E+05 | 0.298E+04 | 0.298E+04 | 42.57 | 0.757E+03 |
| Mean | 0.425E+05 | 0.289E+04 | 0.289E+04 | 41.48 | 0.761E+03 |
| Std. Dev. | 0.261E+04 | 0.269E+03 | 0.269E+03 | 0.604E+01 | 0.123E+03 |

Figure 1. Instron report generated by stress program.

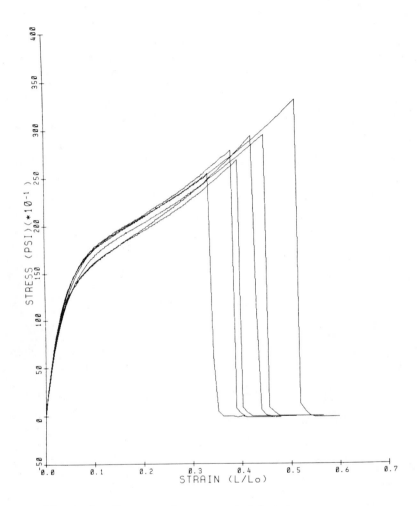

Figure 2. Instron plot generated by stress program.

## Literature Cited

1. Niemann, T. F., Koehler, M. E., and Provder, T., "Microcomputers Used as Laboratory Instrument Controllers and Intelligent Interfaces to a Minicomputer Timesharing System" in "Personal Computers in Chemistry"; Lykos, P., Ed., John Wiley and Sons, New York, 1981.
2. Griffith, A. A., Phil. Trans., A221, 163 (1966).
3. Brown, W. F., and Strawley, J. E., ASTM STP410 (1966).
4. Scott, R. L., Advances in Laboratory Automation - Robotics 1984, Zymark, Hopkinton, MA (1984) p. 151.

RECEIVED November 14, 1985

# 13

# Software for Data Collection and Analysis from a Size-Exclusion Liquid Chromatograph

**John D. Barnes, Brian Dickens, and Frank L. McCrackin**

**Polymers Division, National Bureau of Standards, Gaithersburg, MD 20899**

This paper describes software that is used for data collection and analysis from a size-exclusion liquid chromatograph. The chromatograph is a commercially available instrument that provides on board microprocessor control of the specimen injection functions. We use a commercially available microcomputer as a passive listener connected to the chromatograph output to collect, store, and analyze the data. The data collection and analysis software is written in FORTRAN. Maximum use is made of graphical displays to aid the user's judgment in interpreting the data. All operations are menu drive, so that the user does not need to be familiar with the computer's operating system. Data archiving functions are built in to facilitate after-the-fact retrieval of the data.

Size exclusion liquid chromatography (1) has been widely used to characterize distributions of molecular weights in polymer specimens. This paper describes a package of computer programs for automatic data collection and data reduction in size exclusion liquid chromatography (2). The programs and the environment in which they operate are carefully tailored to emphasize the interaction between the user and his data rather than the interaction between the user and the computer. The system we describe differs from that described by Koehler et al. (3) in that all functions are performed by a stand-alone system.

Although "turnkey" systems for size-exclusion liquid chromatography are commercially available, we chose to develop our own software using a small general purpose data acquisition computer. We feel that the insights gained in this process will lead to improved standardization of size exclusion chromatography and that the resulting flexibility allows us much more latitude in characterizing the complex specimens that result from some of the research that NBS does.

This chapter not subject to U.S. copyright.
Published 1986, American Chemical Society

This paper does not attempt to provide a detailed users guide
for our software, nor does it attempt to describe the programs in
detail. The authors will be happy to provide copies of the programs
and documentation to parties who are interested in technical
details.

Except for certain library routines used to drive the interface
hardware and some system specific routines from the system library,
all of the routines are written in a dialect of FORTRAN IV and they
are sufficiently modular and portable to be adaptable to other
combinations of computers and chromatographs.

Hardware Overview

The programs described in this paper were developed in the Polymers
Division of the National Bureau of Standards for use with a Waters
Model 150C chromatograph (specific products are mentioned in this
document solely for the purpose of specifying experimental
conditions. Such mention does not imply endorsement by the National
Bureau of Standards or that the mentioned products are the best ones
for the purpose) interfaced to a Digital Equipment Corp. MINC-11
computer. Figure 1 is a schematic representation of the hardware
configuration. The computer is a type made for general purpose
laboratory data acquisition. The hardware requirements for the
system could easily be satisfied by a somewhat enhanced, inexpensive
"personal" computer that includes a simple interface for data
collection and adequate disk storage. The interrupt structure of
the computer should, however, permit buffered data acquisition via
the A/D converters. The hardware should be supported by a single
job operating system that supports chaining of tasks and interactive
command language control structures. The programming can be done in
any high level language that provides suitable interaction with the
data acquisition hardware.

In our system the data collection process is essentially a
passive slave to the chromatograph, which is controlled by its own
internal microprocessor. An amplifier matches the voltage output
from the strip chart recorder terminals on the chromatograph to the
A/D converter input. The data collection program uses the
"Equilibration Pulse" and "Injection Pulse" relay closures shown in
Figure 1 to synchronize the data collection process with the
operation of the autosampler on the chromatograph.

The computer stores programs and data on two floppy disks, each
with a capacity of about 500 kbytes. One of the floppy discs
contains the operating system, the needed utility programs, and the
programs for data collection and reduction. The other floppy disc
contains all of the user's data and result files.

A typical set of analyses requires that the user prepare his
samples, perform a setup procedure on both the minicomputer and the
chromatograph, and activate the data collection process.

Software Overview

The menus shown in Figure 2 provide all of the functions that are
required to collect, reduce, and archive the data. The menus, which
are actually programs written for the operating system's command
language interpreter, conduct dialogues with the user that translate

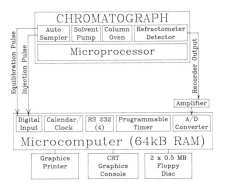

Figure 1. Hardware Block Diagram for NBS size-exclusion liquid Chromatography System.

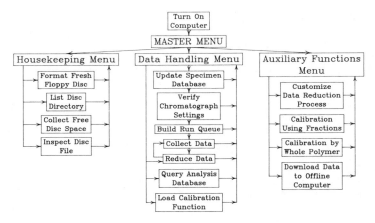

Figure 2. Software Modules for size-exclusion liquid Chromatography

responses to simple English questions into monitor commands for the computer's operating system. This menu-driven mode of operation minimizes the need for users to learn the jargon of the computer's operating system and reduces the number of opportunities for operator error.

The menus form a set of "shells" that envelop all of the individual tasks. Sophisticated users can escape the shell structure imposed by the menu-driven mode of operation and gain added flexibility at the risk of losing part of the protection afforded by the menu-driven procedures.

Some of the tasks performed by the menu selections invoke utility programs that are supplied with the computer's operating system, while other tasks are FORTRAN programs written specifically for our application.

Access to the system is gained through an automatic startup command procedure (the "Turn on Computer" box in Figure 2) which sets the terminal parameters, sets certain operating system flags, and loads the correct date and time from the battery powered calendar/clock. At its completion, the startup command procedure invokes the Master Menu command procedure which provides access to the housekeeping, data handling, and auxiliary function menus. The user can, if necessary, return to the Master Menu at any time by simply turning the computer off for ten seconds or so and turning it back on. The answerback key on the console terminal also invokes the Master Menu if it is pressed while the computer is displaying the input prompt for the keyboard monitor.

The Housekeeping Menu contains functions to format fresh discs to receive data, to verify that there is enough space to contain additional data on a disc, to print the directories of disks, and to print out selected data files. The Collect Free Space function squeezes the blocks containing valid data into the beginning of the disc so that the remaining space is available in one large block. The Inspect Disc File function is sometimes useful in diagnosing problems due to instrument malfunctions or inappropriate instrument settings. All functions that write files to the disc include software write protection to guard against accidental erasure.

The Data Reduction Function provides online reports of analytical results immediately after the analysis is complete, but we also provide direct access to this function for the user who wants to reduce his data in a different way.

The "Query Result Archive" function provides tables of results from recent analyses so that the user can readily see differences between closely related specimens. The "Load Calibration Function" entry in the Data Handling Menu allows the user to change calibration functions to suit his particular needs.

The Auxiliary Functions Menu is used less frequently than the other two menus. The first function creates a disc file that steers the online data reduction process in a manner chosen by the user. Some adaptability is necessary here because procedures for picking baselines, etc. will vary from one group of specimens to another.

The calibration function calculation uses multiple linear regression to obtain hydrodynamic volume as a polynomial function of elution volume for a given column set.

The downloading function is provided so that the user can automatically create disks that can be used to transport information from a data disk to an offline computer for further reduction. This function also writes an extract of the result archive to the output disk for use in updating the comprehensive result archive that is maintained on the remote host. The algorithms used for online data reduction are transportable to any computer that supports FORTRAN IV [which is a subset of FORTRAN 77 (4)] so that users can process their data interactively when they are unable to use the data collection computer.

Data Handling Functions

Figure 2 shows that the path through the Data Handling Menu is quite restricted, in that some functions cannot be performed without first performing others. This is done because the chromatograph analyzes a group of specimens in one continuous operation (batch mode). Each set of data must be associated with pertinent archival data. Because of the passive nature of our data collection process, we force the user to pause so that he can verify that the chromatographic conditions that the data collection and reduction programs will eventually use are the same as those that are manually programmed into the microprocessor on the chromatograph. Figure 3 depicts the relationship between the various data structures and the data data handling tasks that are invoked from the data handling menu (see Figure 2).

The structure of the specimen database is dictated by the fact that the specimen carousel in the chromatograph holds up to 16 samples. The set of analytical parameters associated with each specimen position includes the number of replicate injections, the volume of specimen for each injection, the flow rate of the eluting solvent, the duration of the chromatogram, the detector gain, and various timing parameters. A phantom zeroth specimen position is used to define the analytical parameters for injections not specifically programmed into the microprocessor. The operator must manually enter these parameters into the chromatograph's internal microprocessor in order to analyze the specimens in the carousel.

Our specimen database also contains additional parameters that are used to control the data collection process and to provide archival information to each data file written by the collection process. The console display for editing the specimen database is of the "fill in the form" type and the user revises the parameters for each specimen position (including the zeroth) as required. New parameter values are checked for validity at the time they are entered. All other parameters retain the values they possessed during the previous set of analyses. Thus, only minor changes are needed to program for a set of samples similar to the previous ones. All records in the database can be cleared if the analytical conditions are markedly different.

After the user is satisfied that the specimen database is correct, the updating program generates a printed report that the user should read as he manually programs the chromatograph. Discrepancies between the values in the specimen database and those programmed into the chromatograph can lead to loss or mislabelling of data.

After the user verifies the chromatograph settings the specimen database updating program creates the run queue. The run queue contains one record for each data collection step in the upcoming batch. The run queue building task also checks for enough free disk space to hold the expected data files.

The contents of the run queue are displayed on the printer so that the user can verify that all of the analyses are scheduled to be performed as expected. Once the user affirms the correctness of the run queue, the instrument waits for the first injection pulse from the chromatograph. The instrument then enters a loop in which it collects the data from an analysis, reduces the data, and searches the run queue for the next uncompleted analysis. When all analyses are complete the system returns to the master menu and awaits input from the user.

The data collection program monitors the injection pulse from the chromatograph. After a suitable delay the program triggers a double buffered A/D sweep whose timing is determined by the programmable clock. As data buffers become full they are transferred to the main program for plotting and storage on disc. The data collection process terminates when the control program senses the equilibration pulse from the chromatograph or when the user aborts the process from the console. The console terminal displays the chromatogram during the data collection process. Selected graphs and results are sent to the graphics printer following data collection and data reduction.

Data Reduction Functions

The control path for the data reduction functions is depicted in Figure 4, which is an expanded view of the "Reduce Data" Block in Figure 2. The mainline functions provide a direct path from the raw data to a finished result with a minimum amount of input from the user. The functions in the right hand column are used mainly to make minor changes in the parameters controlling interactive data reduction.

The data reduction program can be driven interactively from the console terminal or in batch mode from a command file created by the customizing function on the auxiliary functions menu.

The major problem in data reduction is to select the relevant range of data from a chromatogram such as the one shown in Figure 5. We do not limit our data collection process to the data that will actually be used to calculate the distribution parameters because we find that values outside of the range used in data reduction help to characterize baseline drift and provide indications of the validity of the results.

We find that baseline drift is small if careful attention is paid to the chromatographic conditions. Under these circumstances we need only define the initial and final elution volumes to be included in the calculation of the molecular weight distribution. The baseline is chosen to be a straight line connecting these two points in the chromatogram. Additional schemes for dealing with the baseline are available for specimens containing large amounts of low molecular weight species. The results obtained for such specimens are, however, dubious because of problems associated with calibration and because of interference from injection artifacts

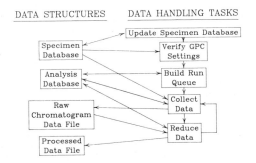

Figure 3. Data Structures and Data Handling Tasks. Arrows connecting tasks with one another represent control flow while arrows connecting tasks and data structures represent data flow.

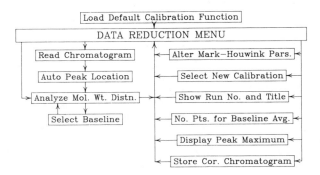

Figure 4. The Data Reduction Process. The arrows depict the control path. Functions in the left column are "Mainstream" functions. The functions in the right hand column are used for custom interactive data reduction.

(the peaks beyond about 52.5 ml in Figure 5). The distribution calculating algorithm provides several different indices that help to assess the quality of the data.

Extracts of significant results from the calculations are stored in the analysis database. These results are adequate for a quick glance at the analytical results. The raw data can be downloaded to an offline computer if further reduction is desired.

## Calibration

Our system provides for several forms of calibration function, but we generally use "universal" calibration (5) and represent the dependence of the logarithm of hydrodynamic volume on retention volume by a polynomial, as in Figure 6. Note that the slope of the function changes dramatically near the ends of the range of applicability. The calibrants at the ends of the range exert a dramatic influence on the form of the fitted polynomial. This behavior demonstrates that the column set must be carefully chosen to fractionate the desired range of molecular sizes.

The scatter of the points from the calibrants in Figure 6 shows that 95% confidence intervals for calculated hydrodynamic volumes amount to +/-10% or so over most of the experimental range.

Determination of the column calibration is greatly facilitated by special computer programs (on the Auxiliary Functions Menu). Peak retention times for calibrating standards are automatically determined during the data reduction process. The resulting values are used as the independent variable to generate a least squares polynomial giving hydrodynamic volume as a function of retention volume for the column set. A similar procedure can be applied when the retention volumes corresponding to quantiles of the cumulative distribution of a specimen with a broad distribution are used as the given quantities. The residual between the given and estimated hydrodynamic volumes are compared to help pinpoint problems with columns or with the standards.

## Future Directions

Experience with the use of our system is driving us in two directions. These approaches have been implemented in new computer programs. The first program is designed to compare chromatograms in terms of hydrodynamic volume rather than elution volume or elution time. This allows comparisons of data from polymer solvent systems in which the Mark-Houwink coefficients are not known, such as polymers undergoing curing. Because the effect of column calibration is removed when the results are expressed as hydrodyanmic volumes, specimens analyzed under differing chromatographic conditions can be compared. This program uses interactive graphic displays for making the comparisons.

The second new program allows the user to compare the shapes of molecular weight distributions. For example, if we have the cumulative distribution of hydrodynamic volume for two polymers we can plot the hydrodynamic volume corresponding to the 10th percentile of the distribution for polymer A against the similarly defined hydrodynamic volume for polymer B. Such a plot, made for the entire distribution of both polymers, is called a "quantile

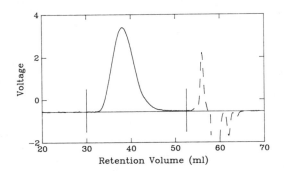

Figure 5. A typical chromatogram. The solid curve is the part of the data that is used to characterize the molecular weight distribution. The dashed portions represent data that are not used. The solid straight line represents the baseline under the chromatographic peak. The vertical lines define the limits of the calibration function for the column set.

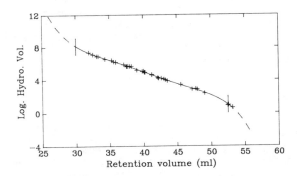

Figure 6. A typical calibration function. The vertical lines delimit the region over which the function is considered reliable. The crosses are the values for various narrow fractions.

plot". The second distribution function can be from experimental data, from a theoretical calculation, or from a reference statistical distribution.

Quantile plots are useful for comparing an observed distribution with a reference distribution, for comparing two observed distributions with one another, for comparing the same data analyzed with two different column calibration functions, or for adding component distributions to simulate an observed distribution. Since this display emphasizes data in the wings of the distribution is it very useful for assessing the influence of the column calibration function. Quantile plots are also a useful diagnostic tool for assessing the user's choice of baseline.

Conclusions

This presentation demonstrates that a small minicomputer can be used to provide a full range of functions for collection and interactive reduction of data from a size-exclusion liquid chromatograph. A number of different users have collected in excess of 5000 chromatograms using this equipment. The experience gained with this system has influenced our approach to the automation of other analytical instruments. Careful attention to control paths, provision of "user friendly" access to the system functions, and careful management of the data archiving functions are crucial to the success of such efforts.

Source codes for the programs and menus discussed in this paper are available upon request to the authors.

Literature Cited

1. Yau, W. W., Kirkland, J. J. and Bly, D. D. "Modern Size-Exclusion Liquid Chromatography"; John Wiley and Sons, New York, 1979.
2. Barnes, J. D., Dickens, B. and McCrackin, F. L. "User's Guide for Size-Exclusion Liquid Chromatography"; NBS Technical Note, To be Published.
3. Koehler, M. E., Kahl, A. F., Niemann, T. F., Kuo, C., and Provder, T.; in "Size Exclusion Chromatography", Theodore Provder, Ed.; ACS Symposium Series 245, Amercian Chemical Society, Washington, DC 1984.
4. "American National Standard Programming Language FORTRAN", Standard X3.9-1978, ANSI, New York, 1978.
5. "Standard Test Method for Molecular Weight Averages and Molecular Weight Distribution of Certain Polymers by Liquid Size-Exclusion Chromatography (Gel Permeation Chromatograph - GPC) Using Universal Calibration"; ASTM D-3593-80, Annual Book of ASTM Standards, ASTM, Philadelphia, PA, revised annually.

RECEIVED February 25, 1986

# 14

## An Automated Apparatus for X-ray Pole Figure Studies of Polymers

**John D. Barnes[1] and E. S. Clark[2]**

[1] Polymers Division, National Bureau of Standards, Gaithersburg, MD 20899
[2] Department of Chemical, Metallurgical and Polymer Engineering, University of Tennessee, Knoxville, TN 37996

> We have adapted a commercially available x-ray diffractometer normally used for structure determinations on single crystals to operate as a very flexible device for performing x-ray pole figure determinations and related studies on polymeric materials. Descriptions of crystallite orientations, as provided by pole figures, are useful in studying many aspects of the behavior of products made from semicrystalline polymers. This paper describes the software that we have written for our pole figure facility. Except for some vendor-provided routines to drive the hardware interface all of our software is written in FORTRAN. Menu driven operation is provided to maximize user convenience.

Processes such as film extrusion, fiber spinning, injection molding, and drawing tend to impart orientation to products made from semicrystalline polymers. Mechanical, dielectric, and optical properties, to mention only three, are often strongly influenced by orientation. X-ray diffraction offers a direct approach to studying crystallite orientation because the intensity that is diffracted into a detector placed at an appropriate position is directly proportional to the number of crystal lattice planes that are in the correct orientation for diffraction. The principles of such measurements are well described in textbooks (<u>1</u>,<u>2</u>).

This paper describes a package of computer programs that we have written to provide menu-driven operation of our facility for x-ray orientation studies. This paper describes the overall structure of the package in the hope that our design approach will be useful to future designers of laboratory data collection systems. Persons who are interested in implementing similar systems can obtain copies of the software described herein from the authors.

0097-6156/86/0313-0140$06.00/0
© 1986 American Chemical Society

## Hardware

An apparatus for characterizing orientation must, as a minimum, provide a means of defining the diffraction geometry of an x-ray beam, a means of supporting a specimen in such a way that any direction in the specimen can be aligned with the diffraction vector, and a means of measuring the intensity of the diffracted x-rays. In order to obtain maximum flexibility we chose to use hardware that is sold commercially for structure determinations on single crystals. This hardware configuration is illustrated in Figure 1. The system is a modern version of the one described by Wilson and Clark (3). This configuration overcomes the inflexibility that is inherent in traditional approaches to pole figure data collection, which use various clever mechanical linkages to drive the specimen orientation along a predetermined path while recording the diffracted intensity.

X-ray and neutron diffractometers were among the earliest laboratory instruments to be completely automated (4) and control techniques for them are well understood (5). Advances in computer technology have allowed economical stand-alone facilities like ours to possess a broad range of control and data reduction capabilities. The 16-bit microcomputer supports a versatile set of operating system utilities, including a FORTRAN compiler and other tools for complete program development. We chose 8 inch floppy discs as a storage medium for compatability with other microcomputers in our laboratory. The diffractometer operates as a standalone instrument, with all data logging and instrument control functions performed by the microcomputer. A 32-bit minicomputer in an adjacent room is used for all modelling and most data display functions so that the diffractometer is available for data collection almost all of the time.

Stepping motors drive the circles of the four-circle goniometer to achieve any desired specimen orientation and diffraction geometry. Two of the angles of the goniometer control the diffraction geometry, these are the $2\theta$ circle and the $\omega$ (or $\theta$) circle. The $2\theta$ circle positions the arm holding the detector so that axis of the detector collimator makes an angle of $2\theta$ with the axis of the x-ray beam collimator (the beam axis). The plane defined by the beam axis and the detector axis (the diffraction plane) is horizontal. The $\omega$ circle positions the mount for the $\sigma$ and $\chi$ circles so that the normal to the $\chi$ circle makes an angle of $\theta$ with respect to the beam axis. The $\phi$ axis of rotation (spindle axis) rides on the $\chi$ circle in such a manner that the spindle axis lies in the plane of the $\chi$ circle and makes an angle $\chi$ with respect the vertical. It is evident that any vector representing a direction in the specimen can be rotated into the plane of the $\chi$ circle by an appropriate rotation about the spindle axis. This vector can then be made to lie in the diffraction plane by a rotation of the $\chi$ circle.

Since the $\omega$ angle tracks the $2\theta$ value, we need specify only 2 $\theta$, $\phi$ and $\chi$ in order to correctly label the intensities read by the detector. In orientation texture studies the diffracted intensity is mapped as a function of $\sigma$ and $\chi$ for a fixed $2\theta$. This provides (under appropriate assumptions) a measure of the probability distribution for the normals to a specific set of diffracting planes for the crystallites that make up the specimen. We will show examples of such intensity maps in later sections of this paper.

The detector signal is conditioned through a single channel pulse height analyzer whose output pulses are fed to a scaler-timer in the single crystal controller.

The computer and the single crystal controller communicate data and instructions through an interface card on the the internal bus of the minicomputer. The data and control registers of this interface are mapped directly into the memory space of the minicomputer. The single crystal interface board also provides interrupts that are used to synchronize events in real time with the driving software. The vendor of the single crystal interface provides a library of FORTRAN callable subprograms to drive stepping motors, collect counts, actuate relays and detect collisions between the collimators and the goniometer circles.

Note that the graphics printer is wired into the terminal so that graphics displays on the CRT screen can be dumped directly to the printer. One of the two floppy disc drives is reserved for the minicomputer operating system and the control programs. The other disc drive is used for the user's data. Data files from the user's floppy disc can be read directly by the 32-bit minicomputer.

The specimen is usually in the form of a rectangular prism approximately 1 mm on a side and 6 to 10 mm long that is mounted in a standard single crystal goniometer head. Since "grain sizes" in polymers are usually very small, and since polymers are usually weak absorbers of x-rays, we perform pole figure data collection in the transmission mode using a small specimen, thus eliminating the need to average over a large area of the specimen while taking data in a reflection mode, as is often required in pole figure studies on absorbing materials.

We have not found the restriction to the transmission method to be a problem because it is easy to make specimen preparation jigs for assembling stacks of film and bundles of fibers into the rectangular prism form. Specimens from sheets or other engineering shapes can be prepared by cutting.

Software Overview

We use a single job operating system. The menus shown in Figure 2 provide all of the functions that are required to collect, display, and archive the data. The menus, which are actually programs written for the operating system's command language interpreter, conduct dialogues with the user that translate responses to simple English questions into monitor commands for the computer's operating system. This menu-driven mode of operation minimizes the need for instrument users to learn the jargon of the computer's operating system and reduces the number of opportunities for operator error.

Sophisticated users can escape the shell structure imposed by the menu-driven mode of operation and gain added flexibility at the risk of losing part of the protection afforded by the menu-driven procedures.

Some of the tasks performed by the menu selections invoke utility programs that are supplied with the computer's operating system, while other tasks are FORTRAN programs written specifically for our application.

Access to the system is gained through an automatic startup command procedure ("Turn on Computer" in Figure 2) that sets the terminal parameters, sets certain operating system flags, and loads the correct date and time from the battery powered calendar/clock. At its completion, the startup command procedure invokes the Master Menu command procedure. The Master Menu is simply a gateway to the subsidiary menus.

The functions in the "Housekeeping Menu" invoke utility programs in the operating system to allow the user to maintain his data discs in an orderly fashion.

The selections in the "Auxiliary Functions Menu" do not produce archivable data, but are used as aids in setting up a specimen and diagnosing problems with the diffractometer.

We have also set up a software library on the 32-bit minicomputer that allows us simulate the diffractometer operation as an aid to developing and debugging programs offline.

Data Handling

The data structures and the control flow through the data handling activities are depicted in Figure 3. We use a batch approach to data collection so that the instrument can perform a series of data acquisition tasks without requiring operator intervention. A batch of tasks may take several days when diffraction is weak or when there are interferences. The structure of the data handling activities is governed by the batch nature of the data collection process.

Information regarding specimens and tasks is stored in two databases. The specimen database contains descriptive information and operating parameters for a number of specimens. The run database contains information regarding the data collection activities that have been performed or that are pending.

We force the user to review and, if necessary, revise the information in the specimen database prior to any data collection activities. The updating function for the specimen database is a menu-driven editor in which the user selects parameters that need to be changed. Data from an earlier specimen can be copied into the entry for a later specimen, thus reducing the number of keystrokes required when running similar specimens.

The information stored in the specimen database is sufficient to identify the particular specimen and the material from which it is made. Other parameters provide information on the orientation of the specimen and on its unit cell parameters. These latter parameters are used by the data collection tasks and the crystal geometry calculation function to determine diffraction angles, the angles between crystal planes, etc. The user can store information on several specimens in the specimen database, thus permitting him to easily remount and rerun a specimen after looking at the collected data.

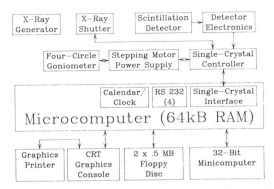

Figure 1. Hardware block diagram for NBS Polymer Pole Figure Facility.

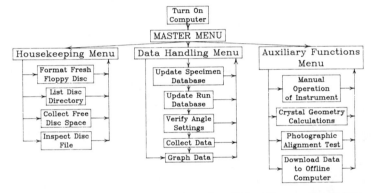

Figure 2. Software modules for Pole Figure Facility.

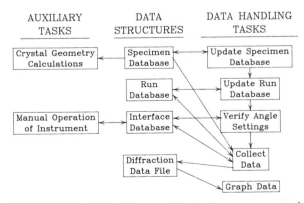

Figure 3. Data Handling Functions. The arrows connecting data structures with tasks indicate information flow. The arrows connecting tasks indicate control flow.

Only one specimen is flagged as being mounted on the machine at any given time. Each data collection task passes the archival information for the mounted specimen to the data file that it creates.

The updating function for the run database is another editing program that is used to create an entry in the run database for each data collection task in a batch. The run database entry contains optional descriptive information as well as the parameters required to drive the task. The user edits the parameter values to suit his needs for the tasks to be performed. Parameters in the run database are validated before the run entry is made permanent. Flags in the run database entries indicate whether the associated task has been completed or is still pending.

To illustrate the kinds of parameters that define a typical data collection task, consider the example of a pole figure. The user first selects a diffraction angle that corresponds to a particular reflection in the unit cell of the specimen. The user must specify the length of time for counting at this $2\theta$ value. At his option the user may select $2\theta$ values near the main peak to be used to collect data for background corrections. Counting intervals for these must also be specified. The range of $\chi$ and $\phi$ values over which data are to be collected must also be specified, together with the number of points to be used to span the ranges. Brief remarks pertaining to the particular data collection task can also be entered into the run database for later reference.

A data collection task updates the run database by associating a run number with the entry for the task and by setting the completion status in the entry. The run number is also used to assign a name to the data file created by the data collection task, thus guaranteeing that each data file has unique name.

We provide a query capability for the run database so that the user can review his task assignments and list the run numbers that are associated with the data files created by the data collection task.

Once the user is satisfied with the task queue that he has constructed by updating the run database, he must ensure that the angle readings on the goniometer circles are correct. These values are contained in the interface database.

The interface database is a small block of memory that is maintained as long as the computer is turned on. The interface database is initialized as part of the startup command procedure. Values for the diffractometer angles are obtained from a disc file that is periodically revised by the application programs. The values may, however, be incorrect because the user may have decoupled the diffractometer circles from their stepping motor drives as part of an alignment procedure or to resolve a collision. The interface database may also be corrupted by an abnormal program termination or a power failure. We provide a simple function for updating the angle values in the interface database.

Each data collection task creates a file of output data (the diffraction data file in Figure 3). The file header contains information from the specimen database and parameters set by the data collection task using the data in the run database. The diffraction data files are, therefore, completely self-contained. We use formatted sequential files because these are the simplest kind to download for offline processing. They are also the easiest kind for other programs to read. The data file header information is used by other programs for use in labelling graphs, etc. The extra space consumed by these files on the floppy disc storage medium has not been found to be a problem.

Data collection is steered by a program that searches the run database for the first pending task. The parameters for this task are then loaded and the appropriate data collection process is invoked. When the task is complete the task manager looks for further pending tasks. This loop continues until there are no more tasks pending or until some abnormal event (such as a collision) terminates data collection.

The data collection task also displays a simple bar graph of the data as it is being collected so that the user can judge whether the data collection process is proceeding as planned.

Data Collection Tasks

We have used the term "data collection task" in a very general context up to now. The user can choose among seven different specific data collection tasks as shown in Table I. Each data collection task uses a different subset of the available control parameters. The tasks represent different ways of mapping the diffracted x-ray intensity. The Standard Pole Figure, the Area Scan, the $\phi$ Scan, the $\chi$ scan, and the Maximum Search map diffracted intensity for a particular reflection (fixed $2\theta$ value) as a function of specimen orientation. In the $2\theta$ scan diffracted intensity is measured as a function of $2\theta$ for a fixed specimen orientation. In the Layer line scan the $2\theta$ angle is stepped between selected limits and the $\chi$ angle tracks the $2\theta$ angle so as to trace out a path along a fixed radial direction for a selected value of z in cylindrical reciprocal space coordinates ($\rho$, $\phi$, z).

The user has no control over the size of the $\chi$, $\phi$ grid that is sampled in the Standard Pole Figure data collection task. The default angular ranges and step sizes have been found to be convenient for many applications. The data in Figure 4 were obtained using the Standard Pole Figure data collection function.

The Area Scan function has been provided for occasions where the Standard Pole Figure function is inadequate for obtaining a good representation of orientation distributions. The mesh size and the extent of the $\chi$, $\phi$ grid can be set to any values that the user desires. For purposes of data smoothing it is undesirable to have large values on the boundary of the grid. The Area Scan function allows the user to specify his grid in such a way as to avoid problems of this kind. The data plotted in Figures 5 and 6 were obtained using the Area Scan function.

Many features of the orientation distribution can be defined by a scan along a line in the pole figure. The $\chi$ scan function allows the user to obtain data along any meridian (fixed $\phi$ angle) of the pole figure while the $\phi$ Scan function collects data along any parallel of latitude (fixed $\chi$ angle). Figure 7 gives examples of $\chi$ and $\phi$ scans. These scans allow data collection in small angular steps without unduly prolonging the data collection process. The full width at half-maximum and the maximum intensity of orientation peaks can often be cleanly resolved from single axis scans.

Table I. Available Data Collection Tasks

| Task | 2θ₀ | 2θ₁ | 2θ₂ | Δ2θ | χ₀ | χ₂ | Δχ | φ₀ | φ₂ | Δφ | t₀ | t₁ | t₂ |
|---|---|---|---|---|---|---|---|---|---|---|---|---|---|
| Std. Pole Figure | U | O | O | NA | O | 90 | 15 | -180 | 170 | 10 | U | O | O |
| Area Scan | U | O | O | NA | U | U | U | U | U | U | U | O | O |
| φ Scan | U | O | O | NA | U | NA | NA | U | U | U | U | O | O |
| χ Scan | U | O | O | NA | U | U | U | U | NA | NA | U | O | O |
| Layer Line Scan | U | NA | U | U | U | NA | C | U | NA | NA | U | NA | NA |
| 2-θ Scan | U | NA | U | U | U | NA | NA | U | NA | NA | U | NA | NA |
| Search for Maximum | U | NA | NA | NA | U | U | C | U | U | C | C | NA | NA |

Key to Control Parameters:
2θ₀ - Beginning 2-θ (or fixed 2-θ)
2θ₁ - Low 2-θ for Bkg
2θ₂ - High 2-θ for Bkg or final 2-θ
Δ2θ - Step size in 2-θ
χ₀ - Beginning χ (or fixed χ)
χ₂ - Final χ
Δχ - Step size in χ
φ₀ - Beginning φ (or fixed φ)
φ₂ - Final φ
Δφ - Step size in φ
t₀ - Time to count at angle
t₁ - Time to count low Bkg
t₂ - Time to count high Bkg

Key:
U - User must specify or accept default
O - Optional, undefined unless specified
NA - Not applicable
C - Calculated from other parameters

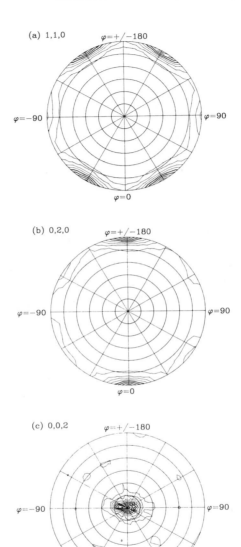

Figure 4. A set of pole figures for three reflections of a highly oriented linear polyethylene specimen. The concentric circles are parallels of latitude, which represent constant values of $\chi$ from 0 deg (outermost circle) to 90 deg (north pole) in 15 deg steps. The straight lines are meridians of longitude, which represent constant values of $\phi$ from -180 degrees to +180 degrees.

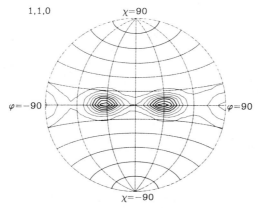

Figure 5. Stereographic projection of data for figure 4a. The point $\phi=0$, $\chi=0$ is at the center of the figure. The meridians from $\phi=-90$ to $\phi=+90$ are the vertical arcs.

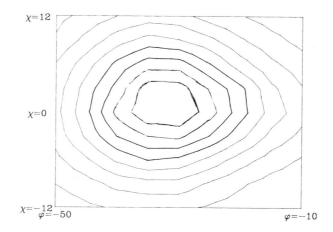

Figure 6. Rectangular contour plot of an area scan from Figure 5.

In some instances the observed intensity over a range of $2\theta$ values around the chosen diffraction peak position contains contributions from background scattering. We provide the control parameters $2\theta_1$, $2\theta_2$, $t_1$, and $t_2$ for use with the first four data collection tasks in Table I so that the user can attempt to compensate for these backgrounds by subtracting intensity values from $2\theta$ settings that are slightly displaced from the value of interest. The values of these angles ($2\theta_1$ and $2\theta_2$) must be specified, as must the counting times ($t_1$ and $t_2$). Setting $t_1$, $t_2$, or both $t_1$ and $t_2$ to non-zero values controls whether the background correction process is invoked. We seldom use this approach because it prolongs data collection and there are often other ways to estimate the background contributions.

The maximum search function is designed to locate intensity maxima within a limited area of $\chi$, $\phi$ space. Such information is important in order to ensure that the specimen is correctly aligned. The user must supply an initial estimate of the peak location and the boundary of the region of interest. Points surrounding this estimate are sampled in a systematic pattern to form a new estimate of the peak position. Several iterations are performed until the statistical uncertainties in the peak location parameters, as determined by a linearized least squares fit to the intensity data, are within bounds that are consistent with their estimated errors.

The $2\theta$ scan and layer line scan functions are provided as an aid to obtaining better estimates of the unit cell parameters for the substance being studied and in assessing contributions from background scattering and overlapping diffraction maxima.

Data Display Functions

At the present time the only online facility that we provide for analyzing the data is the display function, which provides graphical outputs tailored to the data in the diffraction data files. The user specifies the run number for the data he wishes to plot. The display program then loads the header from the appropriate data file and determines what kind of plot to display. The user can dump the resulting plot to the printer to provide a record of his experiment. The "revise file" function can be used to create auxiliary files for plotting selected features of the data.

There are three subroutines for making contour plots using various map projections and a subroutine for making x-y plots. The contour plotting routines are used with the Standard Pole Figure and Area Scan diffraction data files, while the x-y plot is used for data from $\chi$ scans, $\phi$ scans, $2\theta$ scans, and layer line plots.

Figure 4 is a set of contour plots of data derived from three Standard Pole Figure scans on a highly oriented specimen of linear polyethylene. The contour plotting software establishes a set of intensity levels that span the data in the diffraction data file. The contour plotting program then searches each cell on the $\chi$, $\phi$ grid defined by the data to determine, by interpolation, the location of points on the cell edges that should be part of the contour lines for each intensity level. A suitable coordinate transformation then maps the points located in this manner onto the plotting surface where they are joined by straight line segments.

The data shown in Figure 4 are sufficient to map out the orientation distributions for all three crystal axes of this specimen. The implications of the data in these particular pole figures have been discussed elsewhere (6). The diagrams in Figure 4 were drawn using an azimuthal equidistant projection. Figure 5 demonstrates the improved clarity that can be obtained by collecting data as an area scan and plotting it with a stereographic map projection.

We provide rectangular mapping of the $\chi$, $\phi$ grid. This mapping is particularly useful for making precise measurements to locate peak maxima within a small region of $\chi$, $\phi$ space. Figure 6 illustrates this type of contour plot.

Examples of $\chi$ and $\phi$ scans are given in Figure 7. Figure 8 is an example of a $2\theta$ scan.

## Data Processing

We have kept the data collection computer and its software simple and economical. The computing power and peripherals needed to provide high quality plotting and data reduction are available on a 32-bit minicomputer that supports a number of our projects. The floppy discs from the data collection process can be read directly into the data reduction computer. Special menus have been set up to facilitate plotting of the data on a high speed color raster display or on a high-quality multicolor pen plotter.

We have found that data interpretation other than simple plotting from texture studies requires a sophisticated array of software tools. These include least squares curve resolving, non-linear least squares fitting, 2-dimensional data smoothing, numerical quadrature, and high-speed interactive graphics, to mention only a few.

Additional software has been developed to merge data from various data collection steps and to model the data using suitable statistical distribution functions. We are working on software to perform corrections for absorption, specimen shape, and misalignment. Library routines for 2-dimensional data smoothing and integration are being adapted to the calculation of orientation functions and other moments of the probability distributions.

## Extensions to other Computers

All of software described above that is written in higher level languages should be portable to other computers. The difficult part of adapting this scheme lies in devising a library to drive the stepping motors and the counting electronics through a suitable hardware interface, which in turn interacts with the operating system through interrupts, status registers, buffers, and the like. The way in which these tasks are done is, of course, dependent on the host computer. Given the adaptability of modern microprocessors it should be possible to devise more intelligent interfaces that would allow even "personal" computers to perform the kind of data collection described above.

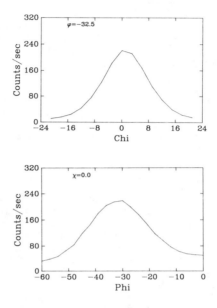

Figure 7.  φ and χ scans for the specimen in figure 4.  These scans are for the (1,1,0) reflection in a region of figure 5 where the intensity is a maximum.

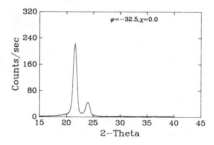

Figure 8.  2 θ Scan for the specimen in figure 4.  This scan was taken at a point where the (1,1,0) intensity is a maximum.

Given that the data collection computer is idle most of the time, it is tempting to consider a multi-user operating system. This would allow more immediate interaction with the data and perhaps permit some program development during data collection. Economic factors to be weighed against this include the need for greatly expanded memory, additional disk capacity, and greater complexity in software maintenance. It is also possible that such a system would "crash" more often than a single job system. We have found that doing software development and data reduction offline, on a much more powerful 32-bit minicomputer that supports a number of other projects, is most sensible for us.

Future Directions

The problem of accounting for backgrounds and contributions from overlapping diffraction maxima will become more vexing as we turn our attention to more complex crystal structures. We are currently studying the use of position-sensitive detectors, which will allow us to collect data over a range of 2-θ values. Desper (7) has described algorithms for separating contributions from overlapping diffraction peaks and continuous backgrounds, and we expect to make these part of the on-line data acquisition process.

We are also developing an improved approach, based on probability theory, for smoothing the observed data and for describing the features in orientation distributions. Since this approach relies heavily on non-linear least squares techniques, it is best done off line.

Most interpretations of orientation texture data are based on the assumption that the specimen is a sphere. A spherical specimen shape is not easily achieved for the kinds of materials we are dealing with and departures from sphericity interfere with the calculation of orientation functions (8). We expect to address this problem by calculating empirical correction functions that can be applied to specimen shapes that closely simulate the ones we use. This problem must be attacked using numerical integration in 3 dimensions, which is a computer-intensive activity.

Summary

The simple system of hardware and software described above provides data for orientation characterization of polymer specimens. The system is used to conduct research on a wide variety of materials rather than to provide analytical results for process control. Given the nature of this usage we have designed the software to emphasize ease of use by operators with a wide range of skills. The operating system environment has been chosen to allow flexible operation of tasks in a menu-driven environment. The task structure and the associated data structures are designed to provide an orderly flow of information and control.

Literature Cited

1. "X-Ray Diffraction Methods in Polymer Science", Alexander, Leroy E.; Wiley-Interscience, New York, 1969.

2. "X-Ray Diffraction by Polymers";Kakudo, Masao, and Kasai, Nobutami; Elsevier, New York; 1972.

3. Wilson, F. C., and Clark, E. S.; "Proceedings of IBM Scientific Computing Symposium on Computers in Chemistry"; International Business Machines Corp, White Plains, NY; 1969.

4. Busing, W. R., Ellison, R. D., Levy, H. A., King, S. P., and Roseberry, R. T.; Report ORNL-4143, Oak Ridge National Laboratory, Oak Ridge, TN; 1968.

5. Clegg, W., "Methods and Applications in Crystallographic Computing", S. R. Hall and T. Ashida, Eds., Clarendon Press, Oxford, 1984.

6. Clark, E. S, Journal of Educational Modules for Materials Science and Engineering, Vol 1, No. 3, Fall 1979.

7. Desper, C. Richard, "Technique for Measuring Orientation in Polymers", AMMRC Report # TR 73-50, Army Materials and Mechanics Research Center, Watertown, MA, 1973.

8. (Translated from German by Peter M. Morris)," Texture Analysis in Materials Science: Mathematical Methods":  Bunge, H.-J.; Butterworths, Boston,  MA, 1982.

RECEIVED May 5, 1986

# 15
# Computers and the Optical Microscope

**M. B. Rhodes and R. P. Nathhorst**

Chemistry Department, University of Massachusetts, Amherst, MA 01003

> The computer age has brought about considerable innovation in the operation of laboratory instrumentation. One consequence of this is the wider acceptance and utilization of the optical microscope as a quantitative analytical instrument. A brief literature survey illustrates the diversity of disciplines and optical methods associated with the development of computer interfaced optical microscopy. This is followed by a description of how our methods of fluorescence, interferometry and stereology, used for characterizing polymeric foams, have incorporated computers.

This presentation is addressed to those individuals who need to characterize or measure systems whose dimensions are found to be within the range covered by any of the techniques of optical microscopy. The experienced optical microscopists will have already either designed and assembled to suit their specific requirements some form of computer interfacing or have at least considered the applicability of some such instrumentation. Therefore to this latter group of individuals, a presentation on the subject of computers and the optical microscope will have limited value. But to all other individuals, especially to those in the materials sciences, it is recommended that they seriously investigate the variety of optical microscopy techniques available and evaluate the increased potential when any degree of automation is incorporated into the operational methods. While this presentation cannot hope to cover every detail or variation of instrumentation and methods it will attempt to introduce the reader to a representative sampling of the extensive and available literature with the hope that interested parties will follow through with their own interests or needs. In addition, this communication will also include a brief description of some of our experiences with optical microscopy and comptuers giving special emphasis to the application of quantitative stereology for foam characterization.

0097-6156/86/0313-0155$06.00/0
© 1986 American Chemical Society

## BACKGROUND

The title of 'Computers and the Optical Microscope' is one that encompasses the extensive field of optical microscope automation, and as such, this history goes back over a great number of years. It is not appropriate for this communication to trace the development of automated optical microscopy, but rather to provide some perspective on the application of various types of relevant instrumentation. Consequently any one interested in characterizing materials can begin to appreciate that many of the available optical microscopy techniques have finally evolved into valid quantitative analytical methods of analysis. The basic instrumentation includes the microscope and appropriate light source, a photomultiplier and/or camera assembly, a real time interface unit, a computer with data storage, a video display and tape recorder, and a printer. This instrumentation, whether integral or modular, will become more detailed when a specific application is envisioned, such as fluorescence or interferometry, but basically the major components are as noted. Automation then can be broken down into three general areas, (a) the control of the microscope itself, such as the stage movement, diaphragm and shutter operation, focusing, etc., (b) the collection and storage of experimental data, such as fluorescent or absorption intensities, and (c) the processing of this experimental data such as for statistical validity, image enhancement, or merely for a graphical presentation. These three areas of automation may function independently of each other or in any combination.

The most obvious place to begin any such discussion of automated optical microscopy is in the area of image analysis. This has long been an active field of development partly because it is inherently broad and thus can be incorporated into so many different investigative scientific disciplines. There are three basic approaches to image analysis, size measurement, image enhancement, and feature or domain distribution. Consequently interest in automating for any of these applications can be found in investigations ranging from haemotology to satellite reconnaissance. Literature references pertaining to the most simple form of automation, that of stage control, began to appear in the 1950s from cytologists like Caspersson (<u>1</u>). Humphries, in his review paper of 1963, addressed the need for automatic measurement from a broad point of view including introductory stereology (<u>2</u>). Wied is the editor of an informative volume on quantative cytology, a publication that not only includes presentations dealing with automation but articles of great interest on optical microscopy per se (<u>3</u>). Innovation in automation reached a high point with the introduction of the Quantimet and while these type of units are currently being replaced with more compact and less expensive instruments, many Quantimets that are still in operation provide satisfactory service in a wide variety of ways (<u>4,5</u>). The rapid development of instrumentation for this type of image analysis is mainly the result of the availability of both computer instrumentation and software. Readers are referred to a recent review by Rosen which covers the electronic instrumentation currently available for optical microscope image analysis (<u>6</u>). This is a particularly informative publication dealing with basic descriptions and problems and including, not only the available instrumentation but, the names and addresses of the manufacturers. Piller is an author whose publications are well worth reading. Although Piller's field is one of photometry, he addresses many issues of interest to anyone who expects to automate the optical microscope (<u>7-9</u>). Although the literature has seen a number of publications dealing with the problems and advantages of incorporating various video techniques for image analysis, two are especially informative with respect to instrumentation. Vrolijk, Pearson and Ploem describe an image processing system initially designed do prescreen data prior to its acceptance by the computer (10). The second report deals with increased contrast enhancement of the video signal by a series of modular units (11). Finally, it is suggested that the reader pursue the collection of references, compiled by Kentgen on Chemical Microscopy, that appeared in the 1984, and especially the 1982, Reviews issue of Analytical Chemistry (12).

A logical extension of image analysis is the need to correlate the specific optical characteristics with the spatial features of the sample that were observed under the microscope. This leads to the subject of microspectrophotometry, which from the point of view of gaining perspective, can include densitometry, inteferometry and fluorescence. Valuable information has appeared from a variety of fields contributed by: the cytologist who needs to determine dry mass from scanning microinterferometric techniques, the forensic scientist who requires precise spectrophotometric identification from paint samples, and by the electronics scientist who is monitoring silicon chips for thermal output with infrared microscopy (13-15). While there exist many valuable papers from a wide selection of applications (and again the reader is advised to consult the Analytical Chemistry Review issues), the authors are calling the reader's attention to just a few references taken from diverse disciplines mainly in an effort to emphasize the value of reading publications from fields other than ones own. Finally in this area of spectrophotometry one finds an especially interesting report by Langer and Frentrup in which they describe automated microscope absorption spectrophotometry for rock forming minerals over the range from 250 to 2000 nm (16).

The microscopy application of fluorescence is one application that has accumulated a considerable portion of the total number of references pertaining to automation. Two papers are very worthwhile for future reference. The first report, by Froot, is one in which the application is more industrially oriented. It describes how the optical microscope system can be under operator control for a single or isolated observation or how the same system can be then applied to the on-line monitoring of organic contaminants on silicon chips (17). The other report, by Smith and coworkers, is one of a research nature in which the investigative optical methods had to be tailored to recording valid low level fluorescence associated with carcinogenic hydrocarbon uptake and localization in cells (18). Both of these have a great deal to offer anyone undertaking fluorescence analysis and the former is a basic presentation for individuals considering on-line quality control by observation with an optical microscope of either a fluorescent or non-fluorescent species.

Interferometric methods initially were automated by the cytologists because sample inhomogeneity of the cell necessitated integrating the optical intensities over a statistical sampling of the cellular domains (19-20). The material scientist does not have this imposed restriction but still requires the very basic advantage of this automation for the ease and speed of measurement. Perry and Robinson have described the interferometric measurement of surface topography of magnetic tapes (21). Their first step was to use the Zeiss Mirau optics to photographically produce the interference fringes and to subsequently have them computer analyzed. They then carried this one step further by eliminating the photographic recording and used the computer directly in the fringe analysis (22). For anyone thinking about automation for interferometric methods, the thesis by Koliopoulos is a valuable reference since the subject matter is very relevant to the automation of the fringe analysis of interferograms (23).

Finally one should not overlook the wealth of material available from the various microscope manufacturers. Such publications describe the instrumentation and, where ever appropriate, provide data relevant to specific applications.

## DEFINING THE REQUIREMENT FOR A COMPUTERIZED OPTICAL MICROSCOPE

Our laboratory is far from realizing the ultimate design in an automated optical microscope. Changes and additions are made step by step, with careful consideration of as many of the advantages and disadvantages as are immediately obvious, but always working towards achieving the type of instrumentation most appropriate to our requirements. The following descriptions illustrate some of our experiences along the route to automation, consistent with our already on-hand or available equipment, our computer facilities and our budgetary restrictions. And while other laboratories may fail to identify exactly with our situation, they still may find some comment in our descriptions that will aid their planning for future investigations.

Experimental Materials. All the data to be presented for these illustrations was obtained from a series of polyurethane foam samples. It is not relevant for this presentation to go into too much detail regarding the exact nature of the samples. It is merely sufficient to state they were from six different formulations, prepared and physically tested for us at an industrial laboratory. After which, our laboratory compiled extensive morphological data on these materials. The major variable in the composition of this series of foam samples is the amount of water added to the stoichiometric mixture. The reaction of the isocyanate with water is critical in determining the final physical properties of the bulk sample; properties that correlate with the characteristic cellular morphology. The concentration of the tin catalyst was an additional variable in the formulation, the effect of which was to influence the polymerization reaction rate. Representative data from portions of this study will illustrate our experiences of incorporating a computer with the operation of the optical microscope.

Basic Characterization of the Foam. Since our laboratory was already engaged in a polymer foam characterization program using the optical microscope, it was logical to assume that introducing some form of automation into our quantitative techniques would be advantageous. Since one aspect of our studies was concerned with the major cellular characteristics of polymeric foams, this seemed to be a convenient place to begin, because this type of characterization found us performing the laborious tasks of counting, measuring, and identifying cellular structures in much the same manner as was done by the botanists some forty years ago. We chose not to automate along these lines as a result of two preliminary but informative tests. The first attempt used the image directly from the microscope and even had the contrast proven to be satisfactory, there was no convenient way to standardize the contrast level from one position of the sample field to another. The modular units that we tried were unable to accomodate the variability and uniqueness of our sample materials. The other attempt, which also failed to satisfy our needs, involved the color enhancement of black and white photographic images. While this method proved very satisfactory in displaying the cellular structure of the foam, it frequently included features of no morphological significance whatsoever, a fact known to only the microscopist who had recorded the original image at the microscope. So it was not until we investigated the use of the stereological methods that we felt we had a valid type of image analysis suited to our type of investigation. Later in this section we will describe our application of stereology to the characterization of foams.

<u>Fluorescence of Polyurethane Foams</u>. It therefore evolved that our intensive introduction to computers and the optical microscope eventually arose from our need to quantitatively measure fluorescence from the polyurethane foams. Qualitative fluorescence observations coupled with previous morphological studies established significant differences in the emission spectra from a series of these foams. The Zeiss ZONAX provided a formal introduction to automatic microspectrophotometry. The ZONAX was a Zeiss modular single-unit computer terminal, which when integrated with the optical microscope provided full control to operate the microscope as a scanning microspectrophotometer. The operations included control of the monochrometer, shutter sequences, stage movement and photodetector gain for fluorescence and spectral analysis in either transmitted or relection modes. Subsequent data analysis was available as graphs and histograms. The major difficulty we encountered was in the very nature of our investigation. We were not immediately concerned with absolute intensity values, since we did not monitor the thickness of the morphological structures nor the density of the cellular matrix material within the structure. We were looking merely for shifts in the maximum wavelength of the emission spectrum. Since the ZONAX software was designed for experiments that used specific fluorochromes with the appropriate standards, the system was not well suited to our less regimented application. We consequently found it more appropriate to use the ZONAX for data point collection under the controlled and documented optical-electronic conditions provided by the ZONAX software and to then transfer this data to a VAX which proceeded to calculate corrected intensities which then provided tables and graphs for comparisons among our specimens. We have recently acquired a PRO 350 with RS/1 software and when the installation of the appropriate interface unit is made, the 350 will not only collect the fluorescent intensities but also perform all the necessary data processing for the statistics and graphics.

<u>Interferograms</u>. Thin film measurements have been a part of our foam characterization for some time because cell window thickness profiles are of major interest in their correlation with the bulk properties of the foam. We adapted, to our optical-computer system, the method described by Akabori and Fujimoto, in which interferograms are recorded at three different wavelengths (<u>24</u>). We transferred the specific wavelength Jamin-Lebedeff interferometric images from our microscope to the VAX and then analyzed for the fringe shift as a function of wavelength. From this information it was possible to plot a thickness profile as a line scan from any direction across the interferogram. The preliminary work with this method has shown it to be time consuming and we have yet to fully

evaluate the method for its applicability to routine measurements. There are a few modifications that would greatly decrease the time investment, the most important being the use of a faster scanning stage. The major advantage of this method is the elimination of the photographic recording process and the investigator's time at the microscope.

Stereology. This particular application of automated optical microscopy should be one of the most interesting to a materials science investigator. The method is well documented in the literature having been long known to metallurgists and recently has been shown to be of value to medical pathologists (25-26). However, it has yet to be fully exploited within the general physical science discipline (27). There is a close relationship between stereology and image analysis since image analysis includes the measurement of geometrical parameters. But in spite of the many variations associated with image analysis, there has continued to exist a need for quantitative measures that relate specific geometrical or morphological features to bulk characteristics of the system. This can be accomplished by the measuring techniques associated with stereology in which a two dimensional feature, such as an area of a specific phase, can be used to determine the volume fraction of that phase. This technique was demonstrated as long ago as 1848 by Delesse (28). These techniques, up to the present, have been performed manually in order to acquire the necessary data, such as tracing the domains prior to some measurement technique. This part of our communcation will describe how stereology has been automated for data collection and how the data can then be treated. Although the illustration will utilize polymeric foams as the subject material under investigation, it should be emphasized that any material with physically recognizable domains can be treated in a similar manner.

(1) Methods of stereology. (a) Instrumentation. The stereological data was acquired and was initially evaluated using new instrumentation from Carl Zeiss Inc. This instrumentation consisted of a unversal microscope, with scanning stage and three axis stage auto focus controller, a solid state camera, special interface cards, and IBM/XT monochrome and SONY RGB monitors. Figure 1 is a schematic diagram showing the relationship between the computer and the optical microscope. The field of interest under the microscope is optimized for direct viewing by whatever type of optics required. This raw data image is then collected by the digital video camera and sent to the monitor, at which time the operator may want to adjust either the field selection or the microscope focus. This image is stored in the computer and the contrast levels for the image are set with the computer. The operator will continue to modify the data associated with the stored image until the monitor displays an image acceptable for the intended data analysis. At this time the computer can store the image on a disc, transport the image to another computer, and/or initiate the stereological measuring program.

(b) Basic theory of stereology. Although stereological methods have been in use for over one hundred and forty years they are unfamiliar to many scientists. The underlying method is one by which a two dimensional image of a specimen with the appropriate statistical and geometrical calculations provides a three dimensional characterization. It should be immediately apparent that this technique is independent of the specific type of material one uses for the subject under the

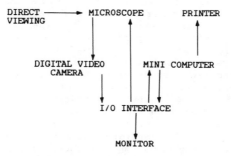

Figure 1. Schematic diagram of the Zeiss microscope-computer instrumentation.

microscope just so long as it possesses phase boundaries recognizable by some optical technique. Two other requirements must be met, one demands the features or domains by distributed homogeneously and the second requires that the sampling be performed randomly.

Stereological methods have often been used without the advantage of a computer or video screen. Such an approach superimposes a grid of dots, lines or areas on the specimen image and counts the inclusions and intersections this grid format shows with the feature of interest within the specimen field. Such a procedure, without automation, is most laborious requiring much effort to establish statistical validity for the measurements.

The present Zeiss instrumentation utilizes the 128 x 128 pixel monitor as a measuring grid, giving a value for $P_T$, the total number of grid points. Four measurements are taken from the image and are used in the subsequent calculations of the selected stereological parameters. These four essential measurements are: $P_H$, the total number of 'hit points' associated with the morphological feature of interest, in this case the gas domain of the foam cell morphology, $P_I$, the total number of intersection points, obtained after the computer has generated a binary image, and $P_{L1}$ and $P_{L2}$, the intersection points associated with the parallel and perpendicular directions of the sample within the microscope field. Table I is a listing of some of the basic stereological parameters with their definitions in terms of the basic measurements (d is the distance between two pixel points on the monitor). Figure 2 is an example of the computer print out arising from the analysis of a foam sample. The programming for these calculations requires a value for the microscope magnification in order to assign a distance between pixels in the image plane. The calculations have followed the procedures established by Underwood (29). The computer will print out each field as it is measured or accumulates data from many fields, until a valid statistical sampling has been made.

TABLE I. BASIC STEREOLOGICAL PARAMETERS

| | |
|---|---|
| VOLUME DENSITY | $V_V = P_H/P_T$ |
| SURFACE DENSITY | $S_V = 2P_I/P_T d$ |
| SPECIFIC SURFACE | $S/V = 2P_I/P_H d$ |
| MEAN INTERCEPT LENGTH | $\overline{L}_3 = 2P_H d/P_I$ |
| MEAN FREE DISTANCE | $\lambda = \overline{L}_3(1 - V_V)/V_V$ |
| MEAN DIAMETER | $\overline{D} = 3P_H d/P_I$ |
| MEAN CURVATURE | $K_m = \pi N_A/2P_I d$ |
| ELONGATION RATIO | $Q = P_{L1}/P_{L2}$ |
| DEGREE OF ORIENTATION | $\Omega_{12} = (P_{L1} - P_{L2})/(P_{L1} + 0.571 P_{L2})$ |

(2) <u>Data collection and analysis</u>. The first data collected on the

15. RHODES AND NATHHORST    *Computers and the Optical Microscope*    163

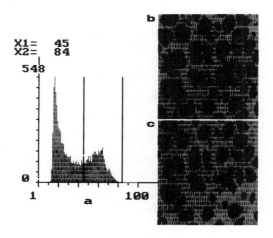

|  | PARAMETERS FIELD # 11 | ACCUMULATION OF 11 FIELDS | STANDARD. DEVIATION |
|---|---|---|---|
| VOLUME DENSITY | .380E+00 | .368E+00 | .993E-02 |
| SURFACE DENSITY | .291E-01 | .269E-01 | .899E-03 |
| SPECIFIC SURFACE | .765E-01 | .730E-01 | .204E-02 |
| MEAN INTERCEPT LENGTH | .523E+02 | .549E+02 | .148E+01 |
| MEAN FREE DISTANCE | .852E+02 | .941E+02 | .404E+01 |
| MEAN DIAMETER | .784E+02 | .823E+02 | .222E+01 |
| NUMBER OF FEATURES(STATISTICAL) | 59 | 572 |  |
| ELONGATION RATIO | .940E+00 | .898E+00 | .326E-01 |
| DEGREE OF ORIENTATION | -.394E-01 | -.698E-01 | .239E-01 |
| MEAN AREA | .481E+04 | .530E+04 | .280E+03 |
| MEAN VOLUME | .377E+06 | .437E+06 | .341E+05 |
| MEAN PERIMETER | .222E+03 | .233E+03 | .626E+01 |
| MEAN SURFACE AREA | .289E+05 | .318E+05 | .168E+04 |
| MEAN CURVATURE | .853E-02 | .814E-02 | .225E-03 |
| NUMERICAL DENSITY | .101E-05 | .848E-06 | .703E-07 |
| EQUIVALENT SPHERE | .896E+02 | .940E+02 | .253E+01 |
| GRAIN SIZE NUMBER | .852E+01 | .838E+01 | .793E-01 |
| MAGNIFICATION | .313E+01 |  |  |

**d**

Figure 2. An example of typical computer print out showing; (a) the histogram of pixel frequency versus intensity with values of maximum and minimum intensity; (b) adjusted raw data image; (c) binary image and (d) calculated stereological parameters.

Zeiss instrumentation was initially intended to be used for an exploratory evaluation and hence no major statistical calculations accompanied these stereological values as they were accumulated. The primary interest at the beginning of the study was to determine the apparent validity of such a characterization method. Consequently samples of a number of different types of foams, polymeric as well as aqueous, were analyzed. Then the correlations between stereological and physical property parameters were assessed from an over-all point of view. The results were sufficiently encouraging that all the preliminary data then became useful for more detailed evaluations. The graphs and statistics that follow here were obtained using RS/1 software with a DEC Professional 350 computer from the initially acquired stereological data. Our initial approach for data presentation, was to plot the average stereological values of a specific formulation against individually selected physical parameters. Figure 3a and 3b illustrate a typical plot of one stereological parameter, MC, the mean curvature, with a selected bulk characteristic of the foam, the 40% compression deflection. Figure 3b illustrates the result of having the RS/1 fit a polynomial to the data. Figure 4 illustrates the type of print out from the PRO 350 that accompanies the statistics and graphing operations. The RS/1 also generates three dimensional graphs. Thus parameters such as mean curvature and surface density can be evaluated for their combined relationship with the compression properties of the foam. Interpretation of such a plot depends on the specific knowledge of how the individual parameters of mean curvature and surface density can be independently controlled by any given formulation. But the ability to easily make such computations and graphics greatly aids in the understanding of foam cell evolution during the polymerization reaction. Quantitative stereology is expected to provide a means by which foams, either solid or aqueous, can be characterized to give valid structure-property relationships, which in turn will allow for more convenient assessment and prediction of a foams performance whether it be physical, chemical, or thermal in nature.

## SUMMARY

The authors hope that readers will consider using both optical microscopy for characterizing their materials and some form of automation with their selected methods. The available optical techniques are numerous and the literature provides a multitude of ideas for interfacing computers. The route selected can be one that uses an already completely designed instrument dedicated to a specific job or one that builds the instrumentation from modular units to satisfy the particular needs of the analysis method. In the first case the investigator must be well aware of what the instrumentation will do and what it will not do as well as being cautious concerning the method by which the software handles the collected data. In the other extreme, the investigator must realize the need for electronics and computer expertise. But in either case, the rewards are worth the investment and within the near future, technology advances will provide almost unlimited versatility in the design of experiments for quantative optical microscopy.

## ACKNOWLEDGMENTS

The authors wish to express their appreciation to Dr. George Reynolds and Robert Heller of the Computer and Information Science Department,

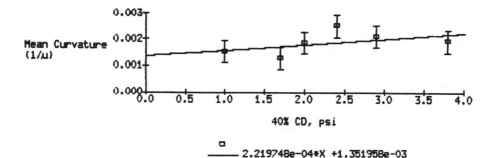

Figure 3a. Typical plot from the RS/1 program showing the linear relationship between Mean Curvature and the tested values for 40% Compression Deflection of the polyurethane foam samples.

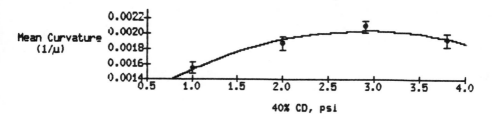

Figure 3b. A plot of the Mean Curvature data versus the 40% Compression Deflection values for the urethane foams made with constant tin catalyst concentration: calculation by RS/1 for a polynomial fit.

STEREOLOGICAL CORRELATIONS FOR URETHANE FOAMS

| 0 PARAMETER | 1 MIL | 2 MFD | 3 ER | 4 MC | 5 40% CD |
|---|---|---|---|---|---|
| 1 MIL | 1.000000 | 0.611290 | -0.003218 | -0.939507 | -0.565933 |
| 2 MFD | 0.611290 | 1.000000 | -0.695369 | -0.575938 | -0.898518 |
| 3 ER | -0.003218 | -0.695369 | 1.000000 | -0.108945 | 0.767748 |
| 4 MC | -0.939507 | -0.575938 | -0.108945 | 1.000000 | 0.502827 |
| 5 40% CD | -0.565933 | -0.898518 | 0.767748 | 0.502827 | 1.000000 |

Figure 4. An example of the RS/1 table of data from a correlation analysis. Column headings are: MC (mean curvature), ER (elongation ratio), MFD (mean free distance between features), MIL (mean intercept length of the features).

University of Massachusetts, and to Dr. Boris Khaykin of Zeiss, Thornwood, New York, for the hours of help and understanding they have already contributed to our optical microscopy endeavors, and finally, to DOW Chemical, Freeport, Texas, for the foam preparation and testing.

## LITERATURE CITED

1. Caspersson, T.; Carlson, L: Svensson, G. 'A Scanning Interference Microscope Arrangement', Experimental Cell Research, 1954, 7, 601-602.
2. Humphries, D.W. 'Mensuration Methods in Optical Microscopy, ADVANCES IN OPTICAL AND ELECTRON MICROSCOPY, Barer, R. and Coslett, V.G., Ed. 1969, 3, 33-95, Academic Press, New York.
3. Wied, G.L. (Editor), Introduction to Quantitative Cytochemistry, 1966, Academic Press, New York.
4. Beadle, C. 'The Quantimet Image Analyzing Computer and its Applications', in ADVANCES IN OPTICAL AND ELECTRON MICROSCOPY, Barer, R. and Cosslet, V.G., Ed. 1971, 4, 361-384, Academic Press, New York.
5. England, B.M.; Mikka, R.A.; Bagnall, E.J. 'Petrographic Characterization of Coal Using Automatic Image Analysis', J. Microscopy, 1979, 116, 329-336.
6. Rosen, D. 'Instruments for Optical Microscope Image Analysis', in ADVANCES IN OPTICAL AND ELECTRON MICROSCOPY, Barer, R. and Cosslett, V.G., Ed., 1984, 9, 323-354, Academic Press, New York.
7. Piller, H. 'Domains of Microscope Photometry in Materials Science', Journal of Microscopy, 1979, 116, 295-310.
8. Piller, H. 'A Universal System for Measuring and Processing and the Characteristic Geometry and Optical Magnitudes of Microscope Objects, in ADVANCES IN OPTICAL AND ELECTRON MICROSCOPY, Barer, R. and Cosslett, V.G., Ed., 1973, 5, 95-114, Academic Press, New York.
9. Piller, H. Microscope Photometry, Springer-Verlag, Berlin, 1977.
10. Vrolijk, J.; Pearson, P.L.; Ploemn, J.S.; A System for the Procesing of Microscope Images, ANALYTICAL AND QUANTITATIVE CYTOLOGY, 1980, 2(1), 41-48.
11. Benton, W.J.; Raney, K.H.; Miller, C.A. 'Enhanced Videomicroscopy of Phase Transitions and Diffusional Phenomena in Oil-Water-Nonionic Surfactant Systems', paper presented at the National AIChE Meeting, March 1985, Houston, Texas.
12. Kentgen, G.A. 'Chemical Microscopy', Anal. Chem., 1982, 54, 244R-265R, and 1984, 56, 69R-83R.
13. Carlson, L. 'A Scanning Recording and Integrating Microinterferometer', Histochemie, 1970, 21, 389-294.
14. Laing, D.K.; Dudley, R.J.; Issacs, M.D.J. 'Colorimetric Measurements of Small Paint Fragments Using Microspectrophotometry', Forensic Science International, 1980, 16, 159-171.
15. Lidback, C.A. 'Scanning Infrared Microscopy Techniques for Semiconductor Thermal Analysis', IEEE Trans. Reliab., 1979, 17, 183-188.
16. Langer, K.; Frentrup, K.R. 'Automated-Microscope-Absorption-Spectrophotometry of Rock-Forming Minerals in the Range 250-2000 nm', J. Microscopy, 1979, 116, 311-320.
17. Froot, H.A. 'The Use of Microfluorescence Analysis for Process Control in the Semiconductor Manufacturing Industry', IEEE Trans. Reliab., 1979, 17, 190-192.
18. Benson, D.M.; Bryan, J.; Plant, A.L.; Gotto, A.M.; Smith, L.C.

'Digital Imaging Fluorescence Microscopy: Spatial Heterogeneity of Photobleaching Rate Constants in Individual Cells', J. Cell. Biol., 1985, 100, 1309-1322.
19. Goldstein, D.J. 'Integrating Microdensitometry', Soc. Exp. Biol., 1980, 3, 117-136.
20. Goldstein, D.J. 'Scanning Microinterferometry', Soc. Exp. Biol., 1980, 3, 137-158.
21. Perry, D.; Robinson, M.; Peterson, R.W. 'Measurement of Surface Topography of Magnetic Recording Materials Through Computer Analyzed Microscopic Interferometry', IEEE Transactions on Magnetics, MAG-19, September, 1983, No. 5.
22. Perry, D. private communication.
23. Koliopoulos, C.L. 'Interferometric Optical Phase Measurement Techniques', Ph.D. Thesis, 1983, The University of Arizona, University Microfilms International, Ann Arbor, Michigan.
24. Akabori, K.; Fujimoto, H. 'A Method for Measuring Cell Membrane Thickness of Polyurethane Foam', International Progress in Urethanes, 1980, 2, 1-61.
25. Gifkins, R.C. 'Optical Microscopy of Metals', Sir Issac Pittman and Sons, Melbourne, 1970.
26. Aherne, W.A.; Dunnill, M.S. 'Morphotometry', Edward Arnold, London, 1982.
27. Gham, J. 'Instruments for Stereometric Analysis with the Microscope - Their Application and Accuracy of Measurement', in ADVANCES IN OPTICAL AND ELECTRON MICROSCOPY, Barer, R. and Cosslett, V.G., Ed., 1973, 5, 115-162, Academic Press, New York.
28. Delesse, A.M. 'Pour Determiner la Composition des Roches', Annales des Mines, 1848, 13, 379-388.
29. Underwood, E.E. 'Quantitative Stereology', Addison-Wesley, Reading, MA 1970.

RECEIVED December 10, 1985

**POLYMERIZATION AND CURE PROCESS MODELING AND CONTROL**

# 16

# Modeling and Simulation Activities in a Large Research and Development Laboratory for Coatings

**D. T. Wu**

Marshall Research and Development Laboratory, E. I. du Pont de Nemours & Company, Inc., Philadelphia, PA 19146

> Microcomputers, introduced in the late 1970's have revolutionized the use of computers. The availability of easy-to-use, inexpensive softwares has also contributed to the upsurge in computer usage. Small systems, with compute power and capability equivalent to large multi-million dollar main frames, are now affordable by small organizations as well as individuals. In this paper the use of computers in applied polymer science will be introduced, using successful applications in our own laboratory as examples. The emphasis is on the application of mathematical modelling and computer simulation techniques. Based on our own laboratory experience, benefits, limitations, and pitfalls to avoid will be outlined, followed by a discussion of factors which in the opinion of the author contribute to our success in applying the modelling and simulation approach.

Microcomputers, introduced in the late 1970's, have revolutionized the use of computers. Small systems, with compute power and capability equivalent to large multi-million dollar main frames, are now affordable by large and small organizations as well as individuals. The availability of easy-to-use, friendly, inexpensive softwares has contributed to the upsurge in computer usage. Years ago, we encountered people who were literally afraid of computers or cynical about the potential capability of the computers. Now, the same people accept computers as a useful tool which can conbribute to their well being. This acceptance of computer is partly due to the availability of well designed, user oriented softwares and partly due to the constant bombardment of success stories in publications and commercials on television. However, if one examines these applications carefully, one will find that these applications are mostly "clerical" in nature. In contrast, this paper will examine the scientific applications of computers in applied polymer science, more specifically, the application of modeling and computer simulation in applied polymer science as related to the coatings industry. The intent of the paper is not to describe the models in detail, but to discuss the types of models being in use, what they do, and benefits derived. Using successful applications in our own laboratory as examples, the advantages, limitations and factors which contribute to successes will be discussed.

## Definition of Modeling and Simulation

In the study of a process or a phenomenon to solve specific problems, mathematical modeling is the process of representing mathematically the essential elements of a process or a phenomenon of the system and the interactions of the elements with one another. Computer simulation is the process of experimenting with the model by using the computer as a tool, i.e. a computer is used to obtain solutions to the mathematical relationships of the model. The model usually is not a complete representation of the system, which often involves inclusion of so many details that one can be overwhelmed by its complexity. Computer is not a required tool to carry out simulation as there are mathematical models which have analytical solutions. Such models can be used to solve problems from a study of the analytical solutions or by simple arithmetic calculations. However, it is most likely that a computer is needed for performing simulations.

## Examples of Modeling and Simulation Activities

The characteristics of the coatings industry are such that most companies market many lines of products and within each line there are many individual products, each containing different ingredients. Usually the revenue from each product or each product line can not justify using this modeling/simulation approach in problem solving, because of the extent of technical efforts required. To be cost effective in our industry, only problems which cut across product lines justify the use of this approach. The two examples discussed here are selected to illustrate the types of problems which are amenable to this approach and the advantages and weaknesses of this approach in R. & D. work.

**Acrylic Polymerization Model.** Acrylic polymers are known to have excellent weathering and functional properties as binders for coatings, and they are widely used in the coatings as well as many other industries. To obtain the desirable property/cost balance, random copolymers instead of blends of homopolymers are frequently used. From the product and process development points of view, problems frequently encountered are:

- Control of exotherm during scale-up.

- Cycle times longer than needed.

- Difficulty in estimating molecular weight and copolymer composition distributions, which affect product performance.

- The need to manufacture the same polymer in reactors of different configurations.

Our acrylic polymerization model was developed to meet the need for solving these problems. Kinetics used are based on fairly well accepted and standard free radical polymerization mechanisms. Although the basic mechanisms are generally agreed on, the difficult part of the model development is to provide the model with the rate constants, physical properties and other model parameters needed for computation. For copolymerizations, there is only meager data available, particularly for cross-termination rate constants and Trommsdorff effects. In the development of our computer model, the considerable data available on relative homopolymerization rates of various monomers, relative propagation rates in copolymerization, and decomposition rates of many initiators were used. They were combined with various assumptions regarding Trommsdorff effects, cross termination constants and initiator efficiencies, to come up with a computer model flexible enough to treat quantitatively the polymerization processes of interest to us.

The model has the capability to estimate a variety of process and product results for many different polymer processes, as shown in Figure 1. It can handle, simultaneously, eight monomers, four initiators and five solvents and chain transfer agents. It has been found to meet the need for solving problems mentioned previously. With well defined goals for polymer properties and process, a technical person can develop the preferred procedure for the synthesis of a polymer with five days of computation efforts and two to five experimental runs. We estimate that without the use of the model, a two to three month effort is needed.

Solvent Formulation System. Another technology area which cuts across product lines is solvent formulation. For each pound of coatings applied, cost of solvents represents an appreciable portion of the total cost. Solvent raw material cost can be as much as one-third of the total raw material cost of coating manufacturers. In addition, solvent formulation affects many end use properties, such as aesthetics, coverage, application latitude, functional properties, etc. For our computerized solvent formulation system, we have developed a number of models which predict properties of solvent blends known to be important in solvent formulation.

    a. Solvent Evaporation Model - This model predicts percent evaporation, solvent composition, and blend solubility parameters, as a function of time under prespecified ambient conditions, i.e. temperature and relative humidity. It is assumed in our model that the controlling resistance to evaporation is the stagnant boundary layer. Non-ideal solution behavior and evaporative cooling are taken into account; polymer-solvent interaction is neglected in the treatment. It is obvious that this model only provides a partial representation of the important factors which affect evaporation; however, our experience has shown that for many problems, the model is sufficiently realistic to provide guidance in formulation.

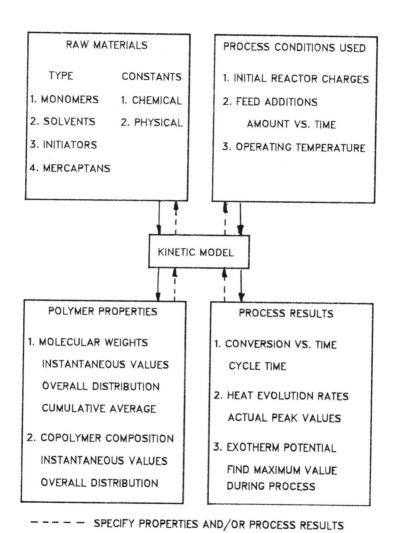

Figure 1. Acrylic Polymerization Computer Program

b. _Determination of Polymer Solubility Envelope_ - One of the most important considerations in the formulation of coatings is whether the polymeric binders are soluble in the solvent blend. Phase behavior of multi-copolymer-multi-solvent system is very complex and cannot be treated quantitatively without extensive experimentation. We have limited our treatment to solubility of one polymer in multiple solvent system. Three dimensional solubility parameter of Hansen (__1-4__) is used to characterize polymer-solvent interaction. Solubility data, expressed as "soluble" or "insoluble" and obtained by dissolving 0.1 gm of polymer in 1.0 gm of solvent, is used to define the solubility envelope in the three dimensional space. We have found the computer program to provide envelopes which are more "consistent", in comparison with those obtained graphically. The computer program stops iteration when it has found the envelope which has the least number of anomalous solvents.

The use of the solubility envelope, together with the volumetric additivity rule for calculating solubility parameters of solvent blend and the solvent evaporation model described previously, allows an approximate assessment whether phase separation will take place or not during solvent evaporation.

c. _Flash Point Model_ - This model predicts flash point of a solvent blend by taking into consideration non-ideal solution behavior, and neglecting the polymer effect. (__5__)

d. _Solvent Blend Viscosity Model_ - Polymer solution viscosity is an important property to consider in solvent formulation. For a given polymer, we need to know both the degree of polymer-solvent interaction and solvent blend viscosity in order to predict polymer solution viscosity. Quantitative estimate of the former is still not possible. We have, therefore, limited our treatment to a prediction of the solvent blend viscosity. Our model is based on Eyring's viscosity equation for liquid mixture with a correction factor (excess free energy of activation) for the free energy of activation of the mixture. The excess free energy of activation is estimated from the excess free energy of mixing using a group contribution method (__6__).

e. _Boiling Point Prediction Model_ - Boiling point of solvent mixture is an important consideration when formulating solvent blends for polymer synthesis under reflux conditions. The model takes into consideration non-ideal solution behavior, but not the polymer effect.

f. _Pure Component Physical Properties Prediction Models_ - Physical properties data for even commonly used solvents are sometimes not available. We have computerized a number of models for prediction of physical properties of our solvents. For each property, there is always a "worst case" prediction model which allows a user to estimate the physical property of interest from chemical structure.

Application of the Solvent Formulation System. In contrast to the acrylic polymerization model discussed previously which is extremely complex mathematically and computation wise, the solvent formulation system is a growing collection of models which are much less complex mathematically and computation wise. However, the system does allow one to evaluate many properties of solvent blend quickly and with relative ease. The system has been found to be valuable in:

- Solvent replacement work.

- Cost optimization.

- Improvement of application properties.

During recent years much technical effort of coating manufacturers has been directed at replacement of solvents due to toxicity or air pollution considerations. Our solvent formulation system has played a major role in these reformulation efforts. We have realized savings both from reduction in technical efforts and in raw materials costs.

From our reformulation experience, we have found that in majority of the cases, computer developed formulations are lower in cost. A 5% reduction in raw material cost can easily be realized.

What is most surprising is the utility of the system in development work to obtain optimum application properties for the product. Our formulators have found that the evaporation curve (% evaporation vs. time) can be manipulated to improve different application properties (e.g., sagging, application lattitude, popping, etc.). Model predictions first provide base points, from which the formulator makes changes in the formulation and correlate the predictions with actual experimental observations. Correlation between model predictions and experimental observations appear to exist for most product lines.

Advantages

Routine Application of Complex Theories. Technical people in the coatings and many other industries tend to shy away from theoretical approaches in their work. There are many legitimate and practical reasons for this, such as complexity of real systems, lack of comprehensive theories, inability to understand theories and the mathematics involved, and time consuming to apply theories. In many cases, it is easier, quicker and more reliable to use the experimental approach. On the other hand, if pertinent theories are computerized and the computer programs contain the required physical properties data, it becomes risk free and easy, to apply theories. We have found that computerization of theories does encourage more people to use them, thus improving both productivity and quality of technical efforts.

Gain Insights Into Problems Impractical to Probe Experimentally.
Models, based on well established mechanisms and "properly" validated
with experimental data, can be useful in probing into areas impractical or impossible to study experimentally. For example, it is very
difficult and time consuming to determine functionality distributions
of oligomers or copolymers; a "validated" polymerization model can
calculate such distributions with relative ease.

Minimize Experimental Work. Most models are not accurate or reliable
enough to arrive at the optimum results a user is looking for.
However, they are usually good enough to eliminate the real bad cases,
allowing the user to concentrate his efforts in more fruitful areas.

Examinations of More Options, and More Optimum Options. The fact
that it is easier to study more cases via computer simulation, the
likelihood of arriving at more optimum options increases.

Means to Correlate and Communicate Technical Information. Instead of
pages and pages of data, graphs, explanations and interpretations,
models can be used to organize data and communicate ideas and thoughts.
It has the additional advantages of lending clarification, permitting
the handling and extrapolation of complex relationships, revealing
gaps and removing ambiguity. It also enhances the probability that
the technical information will be used in future work.

A Good Training Tool. Even though simulators are commonly used in
other industries as training tools, for example, flight simulator,
its potential as a training tool is not as well recognized in coatings field and in applied polymer science. Technical people learn
from experiments and experimental observations. Computer simulations,
replacing actually experiments, greatly facilitate the learning
process, as more "computer" experiments can be carried out in one day
than actual experiments.

Disadvantages

Require Multi-Disciplinary Approach. Model development and implementation of the model for computer simulation require skills and knowledge in a number of scientific disciplines, such as computer programming, applied mathematics, engineering, chemistry and coatings
technology. People with such broad training are scarce and are in
great demand; thus, they are difficult to recruit. It is possible
to use the team approach; however, the project leader must have the
technical breadth and perspective to insure that the model is realistic and practical to use. The technical strength of the project
leader should be in science, engineering, and coatings technology
instead of mathematics and computer programming. Skills in the
latter two fields can be obtained by working with academic, internal
and/or outside consultants.

Long Elapsed Time for Development. With modeling, it has to be
completely developed and implemented before seeing results. It
usually takes one or two man years to develop a model of moderate
complexity. The elapsed time can be shortened through the use of

consultants or licensing of models already developed. In time, more and more computerized models in both the coatings field and applied polymer science will become available.

Availability of Physical Properties Data and Model Parameters. We have found that the development of a data base for physical properties and other model parameters is as time consuming, and intellectually demanding, as the development of the model itself. One will be surprised to know, for example, that vapor pressure data at around 25°C for many commonly used solvents are non-existent.

Approximate Nature of the Results. For practical reasons, models are not developed to represent completely the interrelations of all the variables and to predict the end use properties we look for. Knowing the weaknesses of the model, at times discourages people from using them. Perhaps the strongest argument for using a model which is only a partial representation of reality is that it is better than no representation at all. In addition, our experience has shown that models are most useful if used judiciously and in conjunction with experimental work.

## Factors Which Contribute to Successful Application of the Modeling/Simulation Approach

The modeling/simulation activity has been ongoing in our laboratory for more than a decade. It is now an accepted approach in R.& D. work. Discussed below are the factors, the author believes, which have contributed to our success in introducing modeling/simulation approach in our laboratory:

- Careful selection of areas to work in, from an in-depth understanding of applicable scientific theories, practically in application and what chemists/engineers need.

- Extensive use of internal, external and academic consultants.

- Emphasis on accessibility and "friendliness" of the computer programs.

- Sustained support of programs and data base for physical properties and model parameters.

- Regular training sessions and workshops.

## Acknowledgments

The author would like to acknowledge the help provided by Dr. R. G. Lindsey in the preparation of this paper by sharing his experience in the development and application of the acrylic polymerization model. He would also like to thank the E. I. Du Pont De Nemours & Co. for permission to publish this paper.

Literature Cited

1. Encyclopedia of Chemical Technology (Kirk-Othmer); Mark, H. F., et al.; 2nd ed., 1971; Supplement, pp. 889-910, "Solubility Parameters".

2. Hansen, C. M. J. Paint Technol. 1967, 39 (505), 104

3. Hansen, C. M. J. Paint Technol. 1967, 39 (511), 505

4. Hansen, C. M.; Skaarup, K. J. Paint Technol. 1967, 39 (511), 511

5. Wu, D. T., Finkelman, R., Organic Coatings and Plastics Chemistry Preprints, Vol. No. 38 American Chemical Society: Washington, D. C., 1978: pp. 61-67

6. Wu, D. T., Fluid Phase Equilibria, 1986 (To be published)

RECEIVED December 24, 1985

# 17

# Flexible Control of Laboratory Polymer Reactors by Using Table-Driven Software

**Robert Albrecht-Mallinger**

The Sherwin-Williams Company, Chicago, IL 60628

> The laboratory batch process control system described here balances the needs for power, flexibility, and simplicity by using easily modified configuration tables and a simple reaction control language. Complex operations are built up out of simple commands. Changes to formulations and reactors do not require reprogramming any basic software. The key elements are an architecture of three programs and two configuration files and a formulation processor program that permits the use of macros.

Development of automated batch process control systems has lagged behind that of continuous process control. Flexible factory scale commercial systems have only begun to appear in the last five years (1-4). Increases in the performance/price ratio of small computers are now making automation of laboratory scale batch processes more practical.

These systems must be designed so as to accommodate continual change. They should easily accommodate frequent changes in formulations and reactor hardware. They must be able to control complex operations involving simultaneous tasks. Contrasting with these requirements, the software must be easily maintained, operate reliably, and be easy to use by laboratory personnel.

Sherwin-Williams has developed such a polymer process control system. The methodology used to accommodate the contrasting requirements has two key elements. First, the software is based on a simple architecture that places the definition of changing reactor hardware elements and characteristics in easily modified configuration files (5). Second, the language uses a small number of basic commands to describe formulations and reactor control. Complex operations are described by reference to commands tables (macros) built using several basic commands or other macros.

## Software Architecture

The architecture of the batch control system was designed to let laboratory personnel change formulations and reactor configurations without altering the software. Formula instructions and reactor configurations are stored in tables. The tables are built and used by programs designed to support three types of activities involved in laboratory production of polymers. An Engineer program defines the hardware configuration and sets up instructions on how to perform the important tasks. A Formulator program builds instructions sets describing reactions to be performed. An Operator program executes the instructions. The relationships among these elements are shown in Figure 1.

The Engineer program defines the reactor in dialogue fashion with the user building tables to be used by the Formulator and Operator programs. Among the data and relationships defined are these:
-name of the reactor
-name of each digital input
-name of each analog input, conversion information
-name of each digital output
-name of each analog output, conversion information
-name of each discrete control variable, number of levels, name and digital mask for each
-name of input, output points for heating and cooling control
-name of input, output points for pump control
-tuning constants for heating, cooling, and pumping control
-scan frequency

The resultant table is passed to the Formulator and Operator programs, and becomes the basis of their operations.

The Formulator program guides the user in building an instruction set for performing a polymerization reaction. The user selects commands from key words on a screen menu. The program prompts for required parameter values appropriate for each command. Sets of commands may be grouped together for simultaneous execution. Macro commands, described in the next section, can be inserted and expanded. The program builds an instruction file from the user's responses. The instruction file is passed to the Operator program.

The Operator program controls the production of the polymer and acquires data for later analysis. The operator starts a job by requesting a specific formula and reactor. The tasks required to control the process are initiated, and the reactor put into a ready state. The operator may then begin execution of the formula. A sequencing module steps through the formula, executing in turn each command line. The sequencing module handles the flow control, waiting for completion of each group of commands before proceeding to the next.

There are three CRT graphic displays which the operator may select. They are a picture of the reactor labeled with current process variable values, a trend plot of key variables, and the command file itself, highlighted to show active command lines. The operator can with one keystroke change the console display, go into manual mode, or abort the batch. In manual mode, the reactor

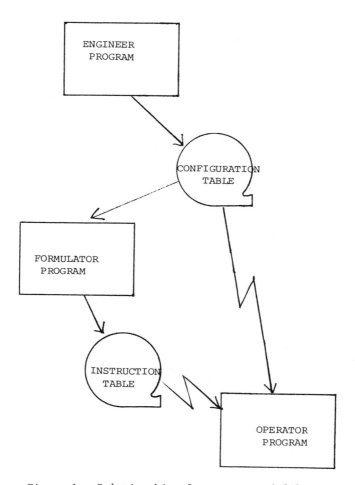

Figure 1.  Relationship of programs and files.

can be directly controlled by the operator using any command available from the reaction control language. The reactor can be returned to automation mode at any time.

Data is recorded in two disk files. All measured process variables are periodically recorded in a trend table. A historical log records the beginning and end of execution of each command, alarms, and notations entered by the operator.

## Reaction Control Language

Overview. At the heart of the automated laboratory polymerization system lies the reaction control language. A standard batch control language has been discussed, but is not yet a reality (6-10). We wanted the formulator to be able to type on a computer terminal a complete description of how to make a polymer using an English-like language with terms he or she is already familiar with. For instance, we have to provide for such actions as initial charging of reagents, controlled pumping of monomers, regulation of heating and cooling of the batch, and operation of agitators and condensers.

Our language includes a set of basic commands which describe fundamental operations such as turning devices on or off, changing setpoints, and operating pumps. The commands are simple English constructions. Simultaneous operations and decision making are handled with "flow control" commands. More complex commands can be built by joining together sets of simpler commands in "macro" files.

Basic commands. There are ten basic commands in the language. They can be categorized in three types: general, time based, and flow control. All formulations are built out of these commands.

General commands initiate some action to be performed immediately, such as setting the state of a discrete control device. Time based commands start processes that require control over a period of time, as in operations that must be repeated at regular intervals. Flow control commands let the control program choose alternative execution paths as needed. They are used to test for the existence of specified process conditions and to change the reaction conditions by branching to other parts of the formula table.

The diagrams in Figure 2 illustrate the syntax of one of each type of the basic commands. Alternative branches in each diagram show variations in syntax that give flexibility to the commands. Figure 3 lists all ten basic commands in somewhat simplified form.

Parallel operation and flow control. Besides the command structure itself, the concept of a group of commands is an important feature of the language. Command sentences from the reaction control language can be grouped together, forming operations that can be performed in parallel (at the same time), or used to define alternative execution paths that can be branched to with IF and GO_TO (two flow control statements used in the language). For instance, the temperature may be held at a fixed setpoint while periodically testing non-volatile weight percent (NVM). If the NVM is low, catalyst is added. If the NVM reaches the specified value, testing is stopped and the program continues with the next command.

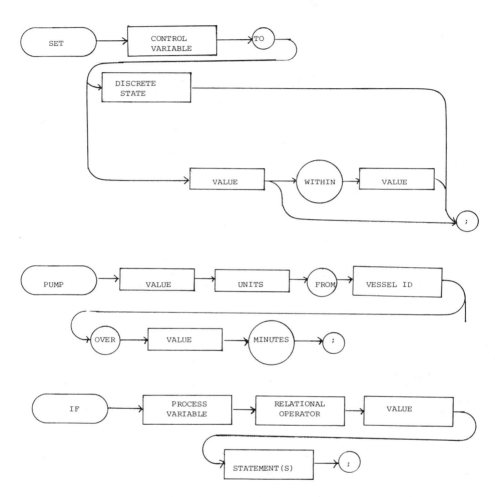

Figure 2. Syntax diagrams.

There are two characteristics of a group that make parallel operation and flow control possible. First, the number of a group gives a reference that can be used in GO_TOs. Second, when time based commands are used in a group, all commands in that group are executed, and remain in effect until all the time based commands finish execution, or the group is terminated by an IF...GO_TO.

Macros. Although the reaction control language resembles English and can be easily read, the description of complex operations can become wordy. Formulators prefer to use their own shorthand in describing a formula. For instance, every chemist has a clear idea of what is meant by "thindown". Having described it once, he or she would like to be able to use a shorthand term to refer to that entire set of commands. Borrowing a term form data processing, this facility is called a "macro". The Formulator program will look up a stored macro table and insert it into the formula. Group numbers are assigned to the added commands, and the user is prompted to fill in the required blanks. In this way, complex commands can be built up without modification to the software.

An example of a macro is shown in Figures 4 and 5. In the first figure we show a formula fragment using only basic commands and a reference to a macro called "THINDOWN". The Formulator program inserts the text from that macro, shown in the next figure. The text from the macro "THINDOWN", shown in the next figure, is inserted by the Formulator program. Suitable group numbers are added, and the chemist is asked to fill in the variable data, such as solvent names and quantities.

Implementation

The programs are now being tested. The specification of the system has already produced benefits by requiring reaction processes to be explicitly defined.

The Engineer and Formulator programs were written in FORTRAN 77 on a Data General MV/8000 running under AOS/VS. The Operator program was written in FORTRAN V on a Data General Model 10/SP using RDOS.

Each program module uses specific system utilities, appropriate to its function, without burdening the other modules with further system dependence. The Operator program needs multitasking and real time capabilities, sensor and actuator I/O, and a graphics CRT. The Formulator program makes heavy use of screen displays to make the formula building process as easy as possible. The Engineer program must contain definitions of the sensor and actuator I/O of each target Operator program.

```
General
        SET <variable> TO <value/state/range>
        LOAD <vessel> WITH <value> <units> <material>
        SAMPLE <variable>
        CALL OPERATOR <text>

Time based
        RAMP TEMP TO <value> OVER <value> <units>
        WAIT <value> <units>
        EVERY <value> <units> <statement #>
        PUMP <value> <units> FROM <vessel> OVER <value> <units>

Flow control
        GO TO <statement #>
        IF <variable> <relational operator> <value> <statement #>
```

Figure 3. Basic commands.

```
        1               SET AGITATION OFF;
        1               SET BLANKET OFF;
        1               SET TEMP_DUTY_CYCLE TO 0.0;
        1               CALL OPERATOR
                            "WEIGH REACTOR, DETERMINE SOLVENT NEEDED
                            FOR _____ % NVM";

        2               SET AGITATION TO LOW;
        2               CALL OPERATOR "ADD _____ TO REACTOR";

        3               SET AGITATION TO HIGH;
        3               WAIT 15 MINUTES;

        INSERT "THINDOWN"

        12              SET AGITATION OFF;

        13              CALL OPERATOR "REACTION COMPLETE";

        14              END;
```

Figure 4. Formula test with reference to macro.

"THINDOWN"

```
6          SET TEMP TO 280 DEGREES;
6          LOAD REACTION WITH 5 GRAMS DI-T-BUTYL PEROXIDE;
6          WAIT 60 MINUTES;

7          SAMPLE %NVM;
7          IF %NVM < 89,
              GO TO 6;

8          SET TEMP TO 250 DEGREES;
8          SET BLANKET OFF;
8          ETC.
```

Figure 5. Expanded macro.

Literature Cited

1. Bansal, S., Advances In Instrumentation, V37 part 2, 694 (1982).
2. Armstrong, W. S. and Coe, B. F., Chem. Eng. Prog., 1, 56 (1983).
3. Gidwani, K. K., Advances In Instrumentation, V37 part 2, 681 (1982).
4. Franey, W. M. and Van Der Jagt, L., Advances in Instrumentation, 669 (1982).
5. Smith, C. L., "Control-Oriented Languages (Table-Driven Software)", pg. 448, "Real-Time Computing", Duncan Mellichamp, ed., Van Nostrand Reinhold (1983).
6. Rosenof, H. P., Chem. Eng. Prog., 9, 59 (1982).
7. Shaw, W. T., Control Eng., 11, 72 (1983).
8. Fihn, S. L. and Nyquist, J. A., Instr. Tech., 10, 57 (1982).
9. Ward, J. C. and Scalera, M. R., Advances In Instrumentation, V37 part 2, 714 (1982).
10. Harrison, T. J., Inst. Tech., 11, 67 (1981).

RECEIVED November 14, 1985

# 18

# Application of State Variable Techniques to the Control of a Polystyrene Reactor

**David J. Hild, Richard E. Gilbert, and Delmar C. Timm**

Department of Chemical Engineering, University of Nebraska, Lincoln, NE 68588-0126

>   This paper extends previous studies on the control of
>   a polystyrene reactor by including (1) a dynamic lag
>   on the manipulated flow rate to improve dynamic
>   decoupling, and (2) pole placement via state variable
>   feedback to improve overall response time. Included
>   from the previous work are optimal allocation of
>   resources and steady state decoupling. Simulations on
>   the non-linear reactor model show that response times
>   can be reduced by a factor of 6 and that for step
>   changes in desired values the dynamic decoupling is
>   very satisfactory.

Timm, Gilbert, Ko, and Simmons (<u>1</u>) presented a dynamic model for an isothermal, continuous, well-mixed polystyrene reactor. This model was in turn based upon the kinetic model developed by Timm and co-workers (<u>2-4</u>) based on steady state data. The process was simulated using the model and a simple steady state optimization and decoupling algorithm was tested. The results showed that steady state decoupling was adequate for molecular weight control, but not for the control of production rate. In the latter case the transient fluctuations were excessive.

The present paper applies state variable techniques of modern control theory to the process. The introduction of a dynamic transfer function to manipulate flow rate removes much of the transient fluctuations in the production rate. Furthermore, state variable feedback with pole placement improves the speed of response by about six times.

<u>Kinetic Model</u>

The kinetic model describes the carbanion polymerization of styrene in a hydrocarbon solvent using n-butyl lithium as the initiator. The mechanism is characterized by four steps:

0097-6156/86/0313-0187$06.00/0
© 1986 American Chemical Society

1) Initiator disassociation
2) Initiation
3) Propagation
4) Polystyryl Anion association.

The kinetic equations describing these four steps have been summarized and discussed in the earlier paper and elsewhere (1,5). They can be combined with conservation laws to yield the following non-linear equations that describe the transient behavior of the reactor. In these equations the units of the state variables T, M, and I are mols/liter, while W is in grams/liter. The quantity A (also mols/liter) represents that portion of the total polymer that is unassociated -- i.e. reactive.

Polymer (mols) $\quad \dot{T} = -QT/V + k_i MI^{.28}$ (1)

Monomer $\quad \dot{M} = Q(M_{in} - M)/V - k_p MA$ (2)

Initiator $\quad \dot{I} = Q(I_{in} - I)/V - k_i MI^{.28}$ (3)

Polymer (mass) $\quad \dot{W} = -QW/V + M_o k_i MI^{.28} + M_o k_p MA$ (4)

Unassociated Polymer $\quad A = ((1 + 4K_{eq} T)^{0.5} - 1)/2K_{eq}$ (5)

where

$$K_{eq} = K^o_{eq} u^{-.2025} \quad \text{(liters/g mole)} \quad (6)$$

$$u = u_o + 2.059 \cdot 10^{-10}(W)^{3.874}/(T)^{1.125} \quad \text{(centipoise)} \quad (7)$$

Each differential equation contains a flow term identified by Q/V (flow rate/reactor volume) and also a reaction term which can be identified by a rate of reaction or equilibrium constant ($k_p$, $K_{eq}$, $k_i$). These reaction and equilibrium constants are functions of temperature which, in this study, was fixed. The viscosity dependence of the equilibrium constant (relating reactive species to total polymer) shown in Equations 6 and 7 was observed experimentally and is known as the Trommsdorf effect (6). Table I lists values and units of all parameters in Equations 1-7.

The isothermal reactor is a system consisting of two physical outputs controllable by manipulation of three inputs. Total volumetric flow rate (Q), inlet initiator concentration ($I_{in}$), and inlet monomer concentration ($M_{in}$) are the three system inputs. Number average molecular weight (MW) and polymer production rate (S) are the chosen outputs which are to be controlled. They are defined by

$$MW = W/T \quad \text{(grams/g mole)} \quad (8)$$

and $\quad S = Q \cdot W \quad \text{(grams/minute)} \quad (9)$

Table I. Parameters Used in This Study

| PARAMETER | VALUE | UNITS |
|---|---|---|
| $k_i$ | .002137 | $liter^{.28}/mol^{.28}/min$ |
| $k_p$ | 12.347 | liter/mol |
| $K_{eq}^o$ | 670.378 | $liter(cp)^{.2025}/mol$ |
| $u_o$ | .66 | centipoise |
| V | 1.523 | liters |
| $M_o$ | 104 | gm/mol |

## Linearized Reactor Model

In order to apply the concepts of modern control theory to this problem it is necessary to linearize Equations 1-9 about some steady state. This steady state is found by setting the time derivatives to zero and solving the resulting system of non-linear algebraic equations, given a set of inputs Q, $I_{in}$, and $M_{in}$. In the vicinity of the chosen steady state, the solution thus obtained is unique. No attempts have been made to determine possible state multiplicities at other operating conditions. Table II lists inputs, state variables, and outputs at steady state. This particular steady state was actually observed by Palsetia (8).

Table II. Steady-State Values Used in This Study

| INPUTS | STATE VARIABLES | OUTPUTS |
|---|---|---|
| Q = .01672 l/min | T = .01896 mol/l | MW = 16172 gm/mol |
| $I_{in}$ = .0259 mol/l | M = .39174 mol/l | S = 5.126 gm/min |
| $M_{in}$ = 3.321 mol/l | I = .00694 mol/l | |
| | W = 306.6 gm/l | |

After all steady state variables are determined, each variable is represented by its steady state value plus a "perturbation" value -- for example $M = M^{ss} + M$. All non-linear terms resulting from this substitution are expanded in a Taylor series about the steady state values and only linear terms are retained. The result of this operation may be summarized by the following matrix equations:

where

$$\dot{\hat{x}} = A\hat{x} + B\hat{u} \quad (10)$$

$$\hat{y} = C\hat{x} + D\hat{u} \quad (11)$$

$$\begin{array}{c}(3\text{x}1)\\ \hat{u} = [\hat{Q}, \hat{I}_{in}, \hat{M}_{in}]^t\end{array} = \text{input variables} \quad (12)$$

$$\begin{array}{c}(4\text{x}1)\\ \hat{x} = [\hat{T}, \hat{M}, \hat{I}, \hat{W}]^t\end{array} = \text{state variables} \quad (13)$$

$$\begin{array}{c}(2\text{x}1)\\ \hat{y} = [\hat{MW}, \hat{S}]^t\end{array} = \text{output variables} \quad (14)$$

$A(4\text{x}4)$, $B(4\text{x}3)$, $C(2\text{x}4)$, and $D(2\text{x}3)$ are constant matrices, and (^) signifies a perturbation variable. A block diagram representation of the linearized system is given in Figure 1.

In this study the linearized equations were used to <u>determine</u> the control strategy, but the non-linear equations were used to <u>test</u> this strategy. For small deviations from the steady state (5% or less) there is very little difference between the responses of the non-linear and the linearized system.

Cost Optimization

This system has three inputs to manipulate, but only two outputs. With such a system, many combinations of **u**'s will produce the same **y**'s in the steady state. Thus, the system is underdetermined. Because of this situation, it is desirable to determine the "best" or "optimum" combination of the inputs that will produce the desired outputs in the steady state condition.

Selection of a unique "optimum" combination must be based on an objective function. The objective function chosen was the "cost of production" function used by Timm, Gilbert, Ko, and Simmons (<u>1</u>).

$$H = \text{Cost}_Q \cdot Q + \text{Cost}_I \cdot I_{in} \cdot Q + \text{Cost}_M \cdot M_{in} \cdot Q \quad (15)$$

where

$H$ = cost of a selected **u** vector    ($/minute)

$\text{Cost}_Q$ = cost of solvent    ($/liter)

$\text{Cost}_I$ = cost of initiator    ($/mole)

$\text{Cost}_M$ = cost of monomer    ($/mole)

Since the system is underdetermined by one variable (3 inputs, 2 outputs), it is possible, for given steady state values of S and MW, to solve the steady state versions of Equations 1 through 9 for $M_{in}^{ss}$ and $I_{in}^{ss}$ in terms of $Q^{ss}$. Then $H^{ss}$ becomes a function of $Q^{ss}$

only and the optimization equation

$$dH^{ss}/dQ^{ss} = 0 \qquad (16)$$

provides a third "output" for the system, giving 3 outputs and 3 inputs. This third output is zero at the steady state. It is monitored, but not used, during the simulations. The effect of this optimization technique is to redefine $\hat{y}$ in Equation 14 as

$$\hat{y} = [\hat{MW}, \hat{S}, d\hat{H}/d\hat{Q}]^t \quad = \text{output variables} \qquad (14a)$$

and to add a third row to matrices C and D to accommodate Equation 16. Since both $\hat{u}$ and $\hat{y}$ are vectors of dimension 3, the system is now "square" and is ready for decoupling.

## Decoupling

For multivariable control systems it is rarely the case that manipulating a single input will cause a change in only one output, leaving the other output(s) unchanged. Changing a single input usually alters all of the outputs significantly.

It is desirable to "decouple" the system so that each manipulated variable appears to affect only one of the output variables. This requires a matrix upstream of the process which, in response to a change in only one of its input signals, will manipulate all the actual process inputs simultaneously so that only the desired output signal changes. If decoupling is to be accomplished only at steady state conditions this matrix is a set of constants. However, if decoupling is required during transient operation the matrix must contain dynamic transfer functions, some of which may not be physically realizable.

In this work only steady state decoupling was used. This is accomplished with the matrix G(3x3), defined by

$$G = [C(-A)^{-1}B + D]^{-1} \qquad (17)$$

The right hand side of Equation (17) is the inverse of the steady state transfer function of the system. Placing the G-matrix ahead of the linearized system, as shown in Figure 2, results in a diagonalized system so that manipulation of any one of the elements of $\hat{r}$ changes only the corresponding element of $\hat{y}$ (at steady state). The input vector $\hat{r}$ no longer corresponds to $\hat{u}$ of Equation (12), but may be thought of as a new set of manipulated variables representing "desired values" of MW, S, and dH/dQ. In practice only the first two of these are manipulated by a controller, since the desired value of dH/dQ remains at zero. Thus only the first two columns of G are actually used.

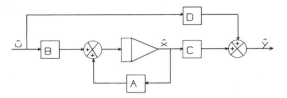

Figure 1. Linear model of the reactor

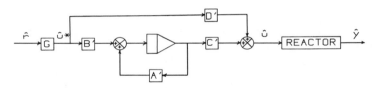

Figure 2. Insertion of dynamic lag for Q

Figure 3 shows some open loop results obtained using the G matrix ahead of the system shown in Figure 1. In Figure (3a) a 5% change in S is called for. The molecular weight (MW) deviates somewhat during the transient period but returns to its initial value. S goes smoothly to its new value after an initial surge. In Figure (3b) a 5% change in MW is called for. During the transient, S varies quite markedly before returning to its steady value. MW is well behaved. Note that the duration of the transient is on the order of 6 hours.

## Effect of First Order Lag in Q

Figure 3 shows clearly that dynamic interactions for this decoupling strategy are minimal for step changes in S, but quite dramatic when step changes in MW are made. The main cause of this dynamic interaction is that S is directly dependent upon the input Q (see Equation 9). When a setpoint change for MW is made, any change in Q results in an immediate change in S. To dampen this effect, a first-order lag was inserted between the flow Q and its input signal Q* with an arbitrarily selected time constant. Figure 2 shows how this modification changes the block diagram. In this figure the block labeled REACTOR consists of everything shown in Figure 1. Matrix D' passes $M_{in}$ and $I_{in}$ directly to the reactor while matrices B' and C' send the Q signal through the first-order lag. Matrix A' contains the time constant for the lag. Matrix G is, of course, the decoupler. This scheme is a form of precompensation similar to that suggested by Silverman ($\underline{7}$).

Figure 4 shows one of the better results obtained through the use of the control scheme illustrated in Figure 2. It is apparent that the large fluctuations in S shown in Figure 3 have been effectively eliminated, although the total transient time has increased. The time constant introduced at A' for this case was 83 minutes.

## Effect of Pole Placement

Figures 3 and 4 were obtained using the steady state defined by the variables and parameters of Tables I and II. Table III shows the poles and time constants computed from the A matrix for this steady state.

The longest time constant is slightly over 1.5 hours. The open loop system requires approximately four times this time constant (6 hours) to reach a new steady state after an upset, as Figure 3 clearly shows. With the added 83 minute time constant for the Q lag described above, this time becomes even longer, as seen in Figure 4.

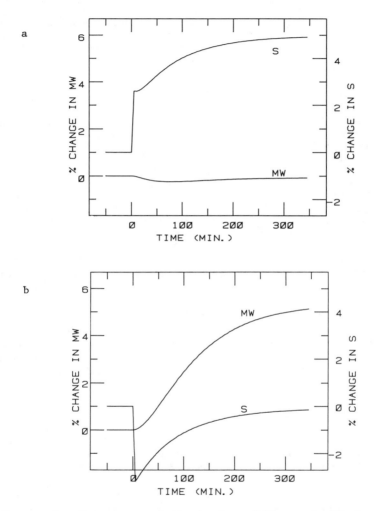

Figure 3. Responses of the system of Figure 1 with decoupling added -- (3a) Change in S; (3b) Change in MW

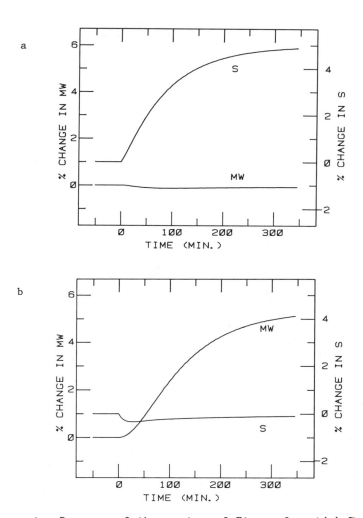

Figure 4. Response of the system of Figure 2 -- 4(a) Change in S; 4(b) Change in MW

Table III. Time Constants for the Linear
System of Figure 1

| POLE | TIME CONSTANT |
|---|---|
| $-0.08231$ min$^{-1}$ | 12.15 min |
| $-0.02670$ " | 37.45 " |
| $-0.01107$ " | 90.37 " |
| $-0.01098$ " | 91.10 " |

As described by Brogan (9) the addition of state variable feedback to the system of Figure 1 results in the control scheme shown in Figure 5. The matrix K has been added. This redefines the input vector as

$$\hat{u} = F\hat{r} - K\hat{x} \qquad (18)$$

and consequently Equations 10 and 11 become

$$\dot{\hat{x}} = [A - BK]\hat{x} + BF\hat{u} \qquad (19)$$

$$\hat{y} = [C - DK]\hat{x} + DF\hat{u} \qquad (20)$$

Whereas the original poles were determined by the matrix A, the poles for Figure 5 are determined by the matrix [A - BK] of Equation 19. Thus, since K is arbitrary, it is possible to modify the system poles as desired. The matrix F corresponds to the G matrix used earlier to decouple the system and, in fact, reduces to G for the case K = 0 (no state variable feedback).

State variable feedback generally has no effect upon the controllability of a system, but can result in a loss of observability (9). Since it is intended to measure state variables directly in the present case, observability is not germane.

Since it was desired not to lose the advantage already gained from using a first-order lag on Q, the scheme shown in Figure 6 was actually used for the pole placement tests. Figure 6 differs from Figure 5 only in that the Q lag of Figure 2 is included.

Hild (5) defined a performance function which enabled him to search for that set of poles which gave the best response for the system of Figure 6. His performance function is

$$P = \frac{\int e_{MW}^2 \, dt}{(MW^{ss})^2} + \frac{\int e_S^2 \, dt}{(S^{ss})^2} \qquad (21)$$

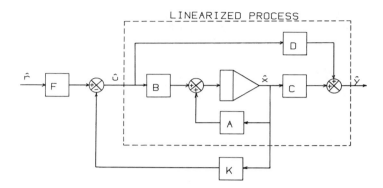

Figure 5. State variable feedback

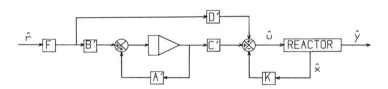

Figure 6. State variable feedback with dynamic lag for Q

The first term represents the integrated squared error in MW when a 5% step change in the S is made holding the desired MW value constant. Likewise, the second term represents the integrated squared error in S when a 5% step change in the MW is made holding the desired value of S constant. Since S and MW have widely different magnitudes, each term is divided by the square of its steady state value to normalize or scale the integrals.

The theoretical limit of optimization would be where P vanishes, which would imply perfect decoupling -- ie. no dynamic interaction between the two output variables.

Starting with an arbitrary set of poles, Hild used Brogan's method (9) to determine the matrices K and F of Figure 6. The integrations of Equation 19 were performed in closed form on the linearized equations and a gradient search was conducted in "pole space" to minimize P. All poles were restricted to negative, real, and distinct values.

Figure 7 shows some of the best results from these pole placement tests. The major difference between these curves and those of Figure 4 is the speed of response. The transient time has been reduced from 6 hours to about 1 hour.

It is of interest to examine how the manipulated variables ($M_{in}$, $I_{in}$, and Q) behave in order to yield results such as those of Figure 7. Unless proper limits are imposed by the model, the manipulations required may be difficult or even impossible to achieve in practice (for example, negative concentrations or flow rates). In this case all three manipulated variables were restricted to positive values. In addition, $M_{in}$ was given an upper bound. No restrictions were placed on rates of change of the variables.

Figure 8 shows the variation of the manipulated variables corresponding to the control problem of Figure 7. Except for some rather abrupt changes in $I_{in}$ and $M_{in}$ at the start of the operation, the functions are smooth and should not be difficult to produce in a real reactor system.

Discussion

A control algorithm has been derived that has improved the dynamic decoupling of the two outputs MW and S while maintaining a minimum "cost of operation" at the steady state. This algorithm combines precompensation on the flow rate to the reactor with state variable feedback to improve the overall speed of response. Although based on the linearized model, the algorithm has been demonstrated to work well for the nonlinear reactor model.

This work covers only open loop response. Use of this algorithm alone on a real reactor would presuppose that the model is very precise -- that a desired MW and S can be obtained merely by

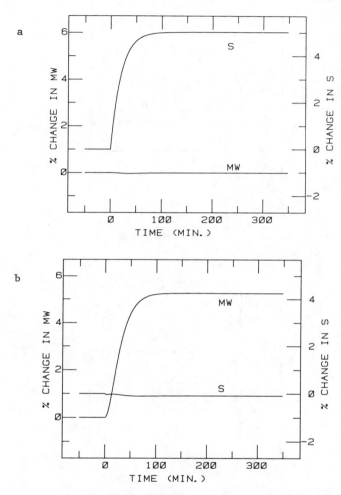

Figure 7. Responses of the system of Figure 6 -- 7(a) Change in S; 7(b) Change in MW

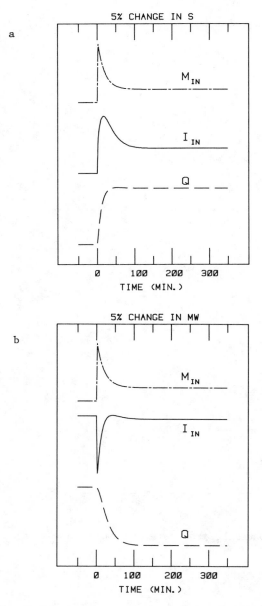

Figure 8. Reactor inputs required to produce Figure 7 -- 8(a) Change in S; 8(b) Change in MW

setting Q, $I_{in}$, and $M_{in}$. This is not likely to be so. In practice, one would place a standard negative feedback loop on each variable (S and MW) between **y** and **r** in Figure 6. These loops would provide final compensation for inadequacies in the model. Because of the decoupling built into the system of Figure 6 each loop would be virtually independent of the other. Negative feedback loops have been added to the model simulator with the expected results (10).

Future Work

The next step in this study is to test this control algorithm on the actual laboratory reactor. The major difficulty is the direct measurement of the state variables in the reactor (T, M, I, W). Proposed strategy is to measure total mols of polymer (T) with visible light absorption and monomer concentration (M) with IR absorption. Initiator concentration (I) can be monitored by titrating the n-butyl lithium with water and detecting the resultant butane gas in a thermal conductivity cell. Finally W can be obtained by refractive index measurements in conjuction with the other three measurements. Preliminary experiments indicate that this strategy will result in fast and accurate measurements of the state vector **x**.

Literature Cited

1. Timm, D. C.; Gilbert, R. E.; Ko, T. T.; Simmons, M. R. In "Computer Applications in Polymer Science"; Provder, T., Ed; ACS SYMPOSIUM SERIES No. 197, American Chemical Society: Washington, D.C., 1981; pp. 3-11.
2. Timm, D. C.; Huang, C.; Palsetia, V. K.; Yu, T. S. In "Polymerization Reactors and Processes"; Henderson, J. N.; Bouton, T. C., Eds.; ACS SYMPOSIUM SERIES No.104, American Chemical Society: Washington, D.C., 1979; pp 376-393.
3. Timm, D. C.; Kubicek, L. F. Chemical Engineering Science, 1974, 29, 2145-2154.
4. Timm, D. C.; Rachow, J. W. In "Chemical Reaction Engineering - II"; Hulburt, H. M., Ed.; ADVANCES IN CHEMISTRY SERIES No. 133, American Chemical Society: Washington, D.C., 1974; p. 122.
5. Hild, D. J. M.S. Thesis, University of Nebraska, Lincoln, 1984
6. Trommsdorf, E.; Kohle, H.; Legally, P. Makromol. Chem., 1947, 1, 169.
7. Silverman, L. M. IEEE Trans. on Auto. Control 1970, 487-489.
8. Palsetia, V. K., M.S. Thesis, University of Nebraska, Lincoln, 1978
9. Brogan, W. L. "Modern Control Theory"; Quantum Publishers, Inc., New York, 1974, pp 52,141,308-312
10. Tatkar, V. K.; Gilbert, R. E.; Timm, D. C. Paper presented to American Control Conference, Boston, June, 1985

RECEIVED December 24, 1985

# 19

# Initiation Reactions and the Modeling of Polymerization Kinetics

**L. H. Garcia-Rubio and J. Mehta**

**College of Engineering, University of South Florida, Tampa, FL 33602**

> Research in the area of Polymerization Reaction Engineering over the last twenty years has resulted in a good phenomenologic understanding of the polymerization kinetics and a series of models capable of describing the reaction behaviour at high conversions. These models contain semiempirical correlations for the rate constants and a number of adjustable parameters. This paper reports on an approach used to obtain explicit expressions for the rate constants and initiator efficiencies as functions of measured conversions, molecular weight averages and the loading of initiator fragments onto the polymer molecules. The expressions obtained, together with experimental data, suggest mathematical representations and correlations for the kinetic parameters. The analysis done yields conditions for the applicability of the classical polymerization equations.

The area of homopolymerization reaction engineering has been extensively studied in recent years (1,9). The main objective in this area has been the development of kinetic models that are capable of explaining the rate behavior and the polymer molecular properties observed during the gel effect. The majority of the models proposed to date are based on the stationary state hypothesis (SSH), the long chain approximation (LCA), and the assumption of constant initiator efficiency (1-9,19). The SSH and the LCA are generally considered valid even though it is recognized that changes in the reaction environment (ie: viscosity) may affect the dynamic behaviour of the radical population. The effects due to changes in the radical concentration are thus reflected on the observed behaviour of the effective termination, propagation and transfer rate constants. Inspection of the term ($2fk_p/k_t$) in the classical polymerization equations indicates that changes in the initiator efficiency are also reflected on the estimates of both kp and kt. In addition, unaccounted changes in the initiator concentration due to reactions like induced decomposition (10) also affect the initiator efficiency and the observed behavior of the rate constants. It is therefore felt that in order to investigate the nature of the gel effect it is necessary to incorporate the changes in both, the initiator concentration and initiator efficiency into the existing polymerization models. Quantitative determinations of the number of end groups per molecule by radio tracer techniques (11,13), NMR (12) and UV spectroscopy (14) have provided new information regarding the fate of the initiator fragments during the polymerization of vinyl monomers. The data, interpreted as overall rates of decomposition and initiation efficiencies suggests that at least part of the behavior attributed to the rate constants during the polymerization could be explained in terms of the initiation reactions. This paper reports on the development of vinyl polymerization models that are capable of describing both; the initiator behavior observed and the molecular weight development.

Literature Review

Experimental data on initiator efficiencies based on the direct measurement of the products of the initiator reactions is rather scant, particularly at high conversions. The main reasons for this have been the difficulties associated with the analysis of the large numbers of decomposition products known to be present in the reacting mixtures. At low conversions, where the ideal polymerization kinetics are expected to be valid, the experimental approaches have been based on measurements of some of the decomposition by-products, rates of polymerization and on the determination of the chain lengths (15-21). The results obtained have been, in general, contradictory (22). A notable exception have been the semi-empirical models derived by Hamielec et al (18,19,21) that correlate the initiator efficiency to viscosity of the reacting mixture and, the cumulative values of the efficiency obtained by Yenalyev and Melnichenko (21) from conversion mesurements and the corresponding degree of polymerization. The uncertanties in the model structure, however, are included in the effective values of the other rate constants (ie: $kp/kt^{1/2}$). Theoretically, two types of models have been proposed: models based on cage effects (10,18,20) and models without cage effects (19). The cage models suggest that the initiator efficiency is proportional to the monomer concentration and independent of the initiator concentration (15). The second class of models indicates that the initiator efficiency is proportional to both the monomer and the initiator concentration. The later models predict a decrease of the initiator efficiency as function of conversion (19).

$$\frac{f}{fo} = \frac{2}{fo + \left(fo^2 + (Mo/M)^2(C/Co)((2-fo)^2 - fo^2)\right)^{1/2}} \quad (A)$$

where f is the efficiency, C is the initiator concentration, and M is the monomer concentration. The subscript (o) refers to the initial value.

At high conversions, the estimation of the initiator efficiencies have been based on measurements of the addition of initiator fragments onto polymer molecules (11-14). Calculation of the initiator efficiencies using this approach have resulted in abnormal efficiency values (14). The contradictory experimental evidence, and the discrepancy in the theoretical results indicates that additional experimental evidence is required in order to elucidate the behaviour of the initiator efficiency as function of conversion. The required information can be obtained from the rate data and from the instantaneous values of the polymer properties. Knowledge of the instantaneous polymer properties allows the estimation of the instantaneous efficiencies and the point values of the rate constants. These parameter values are then conditional upon the initiation and the polymerization models used. The instantaneous polymer properties can be estimated from existing data by proposing empirical models for the instantaneous properties and then fitting these models to the cummulative data. This approach allows the selection of low order models, avoids some of the known difficulties associated with differenciating experimental data and permits statistical testing for model adequacy.

Experimental Data

Literature data for the suspension polymerization of styrene was selected for the analysis. The data, shown in Table I, includes conversion, number and weight average molecular weights and initiator loadings (14). The empirical models selected to describe the rate and the instantaneous properties are summarized in Table II. In every case the models were shown to be adequate within the limits of the reported experimental error. The experimental and calculated instantaneous values are summarized in Figures (1) and (2). The rate constant for the thermal decomposition of benzoyl peroxide was taken as $\ln k_d = 36.68 - 137.48/RT$ kJ/(gmol) (11).

Model Development

The accepted kinetic scheme for free radical polymerization reactions (equations 1-11) has been used as basis for the development of the mathematical equations for the estimation of both, the efficiencies and the rate constants. Induced decomposition reactions (equations 3 and 10) have been included to generalize the model for initiators such as Benzoyl Peroxide for

Table I. Styrene Suspension Polymerization
90 °C with 2.5% by Weight Benzoyl Peroxide (14)

| Sample | Time Min. | % Conv. | $\bar{M}_w$ x10$^3$ | $\bar{M}_n$ x10$^3$ | Ph-CO$_2$ | End Grp./ Molecule |
|---|---|---|---|---|---|---|
| PS01 | 7 | 7.13 | 29.50 | 15.09 | 1.558 | 1.95 |
| PS02 | 20 | 26.34 | 30.58 | 15.91 | 1.556 | 2.07 |
| PS03 | 36 | 57.47 | 32.28 | 16.46 | 1.457 | 2.04 |
| PS04 | 60 | 72.97 | 35.03 | 17.86 | 1.4156 | 2.09 |
| PS05 | 90 | 79.04 | 37.76 | 18.88 | 1.317 | 2.17 |
| PS06 | 120 | 85.74 | 41.78 | 19.82 | 1.246 | 2.19 |
| PS07 | 155 | 88.84 | 43.53 | 19.45 | 1.293 | 2.22 |
| PS08 | 240 | 91.59 | 46.65 | 21.56 | 1.253 | 2.40 |

Table II. Equations for the Instantaneous Polymer Properties

---

Conversion and Rate of Polymerization

$$X = \theta_1(1 - \exp(-\theta_2(t - \theta_3)))  \qquad \text{(fraction/second)}$$

$$Rp = \frac{\rho(1 + \varepsilon)}{(1 + \varepsilon X)^2} \frac{dX}{dt} \qquad \text{(moles/liter-sec)}$$

$\theta_1 = 2.799$; $\theta_2 = 0.1737\text{E}-03$; $\theta_3 = 316.0$; $\rho$= molar density of styrene
t= reaction time (seconds); $\varepsilon$= volume contraction factor

Number average Molecular Weight

$$Mn = 9520 + \frac{5225}{(1-X)}$$

Weight Average Molecular Weight

$$Mw = 17810 + \frac{10380}{(1-X)}$$

Initiator Loading (grams of initiator fragments/gram of polymer)

$$\gamma = 0.01642 \exp(-0.5576)$$

---

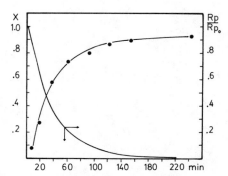

Figure 1. Measured conversions and calculated polymerization rates. (●) experimental values. Solid lines calculated values from equations in Table II.

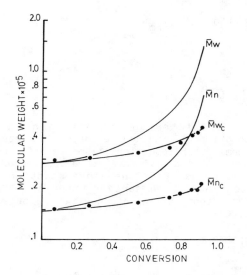

Figure 2. Measured and calculated instantaneous and cummulative molecular weight averages. (●) experimental cummulative values; Solid lines, calculated values (Table II).

which induced decomposition reactions have been reported (10,11,13,14).

Initiation:

$$C \xrightarrow{k_d} 2 R\dot{c}_o \qquad (1)$$

$$R\dot{c}_o \xrightarrow{k_2} R\dot{c}_1 + B \qquad (2)$$

$$R^{\cdot} + C \xrightarrow{k_x} P + R\dot{c}_o \qquad (3)$$

$$R\dot{c}_o + M \xrightarrow{k_{p_1}} R_1^{\cdot} \qquad (4)$$

$$Rc_1 + M \xrightarrow{k_{p_1}} R_1 \qquad (5)$$

Propagation:

$$R\dot{n} + M \xrightarrow{k_p} R\dot{n}+1 \qquad (6)$$

Termination:

$$R\dot{n} + R\dot{m} \xrightarrow{k_{tc}} P_{n+m} \qquad (7)$$

$$R\dot{n} + R\dot{m} \xrightarrow{k_{td}} P_n + P_m \qquad (8)$$

$$R\dot{n} + M \xrightarrow{k_{fm}} P_n + R_1^{\cdot} \qquad (9)$$

$$R\dot{n} + C \xrightarrow{k_x} P_n + R\dot{c}_o \qquad (10)$$

$$R\dot{n} + R\dot{c} \xrightarrow{k_{tc'}} P_n \qquad (11)$$

Depending on the initiator and monomer system secondary decomposition (equation 2), induced decomposition (equations 3,9), primary radical termination (equation 11) or transfer reactions may or may not be important and will have to be considered accordingly in the balance equations. From the above reaction scheme the following equations have been derived under the SSH, the LCA, negligible secondary decomposition and negligible primary radical termination (9,19,20):

$$Rp = \frac{-dM}{dt} = k_p \, M \, R^{\cdot} \qquad (12)$$

$$\bar{r}n \,(t) = \frac{1}{(\tau + \beta/2)} \qquad (13)$$

$$\bar{r}w \,(t) = \frac{2\tau + 3\beta}{(\tau + \beta)^2} \qquad (14)$$

$$\tau = \frac{kfm}{kp} + \frac{kxC}{kpM} + \frac{ktd}{kp^2}\frac{Rp}{M^2} \qquad (15)$$

$$\beta = \frac{ktc}{kp^2}\frac{Rp}{M^2} \qquad (16)$$

In addition to the above equation, in the absence of branching reactions, it is also possible to write a balance for the number of initiator fragments attached to the polymer molecules.

$$\frac{d\rho r}{dt} = (\frac{d\rho r}{dt})_{\text{transfer to polymer}} + 2\,(\frac{d\rho r}{dt})_{\text{term. by comb}} + (\frac{d\rho r}{dt})_{\text{term by disp}} - (\frac{d\rho r}{dt})_{\text{trans to solv}} - (\frac{d\rho r}{dt})_{\text{trans to mon.}}$$

$$\frac{d\rho}{dt}\text{ total} = \sum_{r=1} d\rho r/dt$$

replacing the terms resulting from equations 7 to 10 (6,9,15), for an initiator decomposing into two active fragments, the moles of initiator attached to the polymer molecules can be readily obtained:

$$\frac{dn}{dt} = \frac{1}{2}\left((kx\,C + ktd \sum_{j=1} Rj\,)\sum Rj + ktc \sum_{r=1}^{} \sum_{s=1}^{r-1} R\mathring{s}\,R\mathring{r}\text{-}s\right) \qquad (17)$$

where the indices j, r and s account for monomer units. Equation 17 simplifies to:

$$\frac{dn}{dt} = \frac{1}{2}\,Rp\,(\tau+\beta) \qquad (18)$$

A balance on the number of initiator fragments, or on the end groups per molecule (N(t)) results in:

$$N(t) = \frac{\tau + \beta}{(\tau + \beta/2)} \qquad (19)$$

Alternatively, the grams of initiator bonded per grams of polymer formed Υ (t) can also be readily calculated.

$$Y(t) = (\tau + \beta)\,(\bar{M}f/\bar{M}m) \qquad (20)$$

where $\bar{M}f$ is the molecular weight of the initiator fragments and $\bar{M}m$ is the molecular weight for the monomer. If the transfer reactions (ie: transfer to initiator) imply the addition of a fragment to the growing radical, equation 20 becomes.

$$\gamma(t) = (2\tau + \beta)(\overline{Mf}/\overline{Mm}) \qquad (20a)$$

Based on the literature data available for styrene polymerized with benzoyl peroxide, (10,12,14) transfer to monomer and termination by disproportionation will be neglected. For the initiation step, only primary and induced decomposition reactions will be considered.

## Analysis of the data

From the instantaneous values of the properties reported in Table I, it is possible to determine a maximum of four kinetic parameters. Explicit expressions for the rate constants can be obtained directly from equations 12 to 16 in terms of the parameters $\tau$ and $\beta$, and from these the values of the rate constants can be obtained for a variety of reaction schemes.

$$\beta = (2/\overline{rw})((2 - \overline{rw}/\overline{rn}) + (4 - 2\overline{rw}/\overline{rn}))^{1/2} \qquad (21)$$

$$\tau = 1/\overline{rn} - \beta/2 \qquad (22)$$

where $\overline{rw}$ and $\overline{rn}$ are the instantaneous weight and number average chain lengths.

Note that, for the parameters $\beta$ and $\tau$ to be positive or zero, the instantaneous polydispersity ($\overline{rw}/\overline{rn}$) must be less or equal to 2. This is an important observation because the above condition must be met throughout the conversion trajectory regardless of the actual values of the rate constants. Furthermore, it can be shown that if there is more than one termination mechanism present, the ratio ($\Phi = \tau/\beta$) is finite and it has a value of:

$$\Phi = \frac{-2\left((\overline{rw}/\overline{rn}-2) \pm ((\overline{rw}/\overline{rn}-2)^2 - 4(\overline{rw}/\overline{rn}-2)(\overline{rw}/\overline{rn}-3/2))^{1/2}\right)}{(\overline{rw}/\overline{rn}-2)} \qquad (23)$$

where only the positive root has physical meaning. The discriminant in equation 23 is positive only if

$$3/2 \geq \overline{rw}/\overline{rn} \leq 2 \qquad (24)$$

Therefore, the classical polymerization model is applicable only to those conversion trajectories that yield polydispersities betwen 1.5 and 2 regardless of the mode of termination. Although this is an expected result, it has not been implemented, the high conversion polymerization models reported to date are based on the classical equations for which the constraint given by equation 24 is applicable. The result has been piecewise continuous models, (1-6) where the changes in the parameter values and model structures compensate for this limitation in the classical model. The polystyrene data used here shows cummulative polydispersities which are smaller than 2, therefore, equations 11-20 are considered applicable.

If it is assumed, in accordance to O'Driscoll and White, (10) that the radicals produced from the induced decomposition reaction equation 3 have an efficiency of 1, a balance on the total radical concentration yields

$$R^{\cdot} = \left(\frac{2fk_d\ C}{k_{tc}}\right)^{1/2} \qquad (25)$$

Replacing equation 25 into 11 an expression for the efficiency is obtained.

$$f = \frac{\beta\ R_p}{2k_dC} \qquad (26)$$

A balance on the initiator yields:

$$\frac{-dC}{dt} = k_d\ C + \tau R_p \qquad (27)$$

Equation 27 can be numerically integrated along the conversion trajectory to obtain the initiator concentration as function of time. Therefore, calculation of $\tau$, $\beta$ and C together with the values of M, Rp, rw and rn from the equations in Table II allows the estimation of the ratios $(k_{tc}/k_p^2)$, $(k_x/k_p)$ and the efficiency as functions of conversion. Figure 3 shows the efficiency as function of conversion. Figure 4 shows the variation of the rate constants and efficiencies normalized to their initial values. The values for the ratio $(k_{tc}/k_p^2)/(k_{tc}/k_p^2)_o$ reported by Hui (18) are also shown for comparison. From the definition of efficiency it is possible to derive an equation for the instantaneous loading of initiator fragments,

$$Y' = f(\overline{M_f}/\overline{M_m})\frac{\left(2k_d\ C + \tau R_p\right)}{R_p} \qquad (28)$$

The cummulative values of the initiator loading calculated with equations 20a and 28 should agree, betwen themselves and with the measured values. As it is shown in Figure 5 this is not the case, a correction for the initiator balances seems to be required. Larger efficiency values can be obtained if it is assumed that all the initiator radicals have the same efficiency. Under this condition, the radical concentration is given by

$$R^{\cdot} = \frac{-k_xC(1-f) + \left(k_x^2C^2(1-f)^2 + 4k_{tc}k_d\ C\right)^{1/2}}{2k_{tc}} \qquad (29)$$

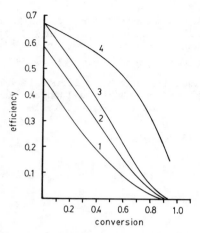

Figure 3. Calculated efficiencies. (1) From the cage effect model and no primary radical termination (Case I); (2) From the assumption of an overall efficiency and no primary radical termination (Case II); (3) From the assumption of an overall efficiency and primary radical termination (Case III); (4) Calculated from equation (A) with fo = 0.663.

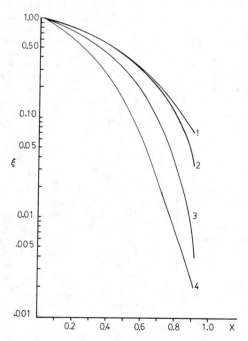

Figure 4. Variation of the rate parameters as functions of conversion: for Case I: (1) $\xi = (k_{tc}/k_p^2)/(k_{tc}/k_p^2)_0$; (2) $\xi = (k_x/k_p)/(k_x/k_p)_0$; (3) $\xi = f/f_0$; (4) $\xi = (k_{tc}/k_p^2)/(k_{tc}/k_p^2)_0$ from reference (18).

which upon substitution into equation 11 gives a new value for the efficiency

$$f = \frac{2Rp(\tau + \beta)}{4kd\ C + \tau Rp} \quad (30)$$

This equation also yields values for $\gamma'$ that are smaller than the experimental values (see Figures 4-5). Finally, if primary radical termination is considered (ie: equation 11), and under the assumption of $(kp_1 = kp)$, a balance on both, macro and primary radicals gives:

$$R\dot{c} = \frac{fC(2kd + kxR^{\cdot})}{kpM - fkxC} \quad (31)$$

$$fC(2kd + kxR\dot{c}) - \left[(1-f)kxC - ktc'R\dot{c}\right]R^{\cdot} - ktcR^{\cdot 2} = 0 \quad (32)$$

Substitution of equation 31 into equation 32 and replacement of R from equation 12 yields, after some algebraic manipulation, the following equations:

$$\tau' = 1/rn - \beta/2 \quad (33)$$

where $\tau'$ now includes the primary radical termination term, ie:

$$\tau' = \tau + \frac{ktc'R\dot{c}}{kpM} \quad (34)$$

$$\tau = \frac{\tau' + \beta + f\psi}{f(\tau' + \beta + 1)} \quad (35)$$

$$\lambda^2 - \lambda(\psi - \tau' + \tau) - f\tau(\tau' - \tau) = 0 \quad (36)$$

$$f = \frac{\tau' - \tau}{\psi\lambda + \tau(\lambda + \tau') - \tau^2} \quad (37)$$

where:

$\psi = 2kd\ C/Rp$ and $\lambda = ktc'Rp/(kp^2M^2)$

The initiator concentration at any time can be obtained from:

$$\frac{-dC}{dt} = kd\ C + \tau Rp\frac{(1 + f\psi + f\tau)}{1 - f\tau} \quad (38)$$

The non-linear equations 21 and equations 35-38 can be solved iteratively to give directly, the instantaneous efficiencies and the the ratios of rate constants (kx/kp), (ktc/kp$^2$) and (ktc'/kp$^2$). The values obtained for the rate constants have been summarized in Table III and in Figures 3 and 6. The results from the calculations show a small difference (ie: less than 1%) betwen ktc and ktc'. Therefore, for all practical purposes they can be considered equal

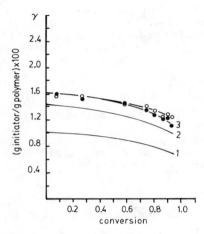

Figure 5. Initiator loadings: (○) Experimental values; (●) Calculated from the molecular weight data and the assumption of two end groups per molecule; (1) Calculated from Case I efficiencies; (2) Calculated from Case II efficiencies; (3) Calculated from Case III efficiencies.

Table III. Summary of Kinetic Parameters

| Equations | $k_{tc}'/k_p^2$ | $f_0$ | $k_{tc}/k_p^2$ | $k_x/k_p$ |
|---|---|---|---|---|
| 21,22,28,29 | 0.453 | 76.39 | 0.451 | – |
| 21,22,29,31 | 0.584 | 76.39 | 0.451 | – |
| 21,34–39 | 0.663 | 76.39 | 0.447 | 76.40 |

at this temperature. The conversion of initiator as function of time is shown in Figure 7. The large discrepancy observed between the initiator consumption calculated from the first order decomposition and the one calculated from equation 37 accounts for the abnormal behaviour of the initiator efficiency obtained from uv measurements (14). The initiator loadings calculated from the efficiency values correspond to those calculated from the Mn's and the assumption of two end groups per molecule Figure 5. It is noteworthy that the ratios of rate constants vary only slightly for the different equations resulting from the assumptions made regarding the efficiency and the initiator balances (see Table III and Figures 4 and 6). However, equations 34-39 are the ones that give the closest agreement between the measured and calculated initiator loadings. If it is assumed that all the chains are linear (and there is no evidence to the contrary for polystyrene synthesized at 90° C), then it must be concluded that a fraction of the primary radicals react with the polymer chains to yield excess ester groups (11,12,14).

The classical polymerization equations, applied to the polystyrene data over the complete conversion trajectory, suggest that all the rate parameters, including the efficiency, vary considerably from the onset of the reaction. The variation of the rate parameters is function of the molecular weight as well as of the concentration of polymer in the reacting mixture (1-8,21). The agreement with the accepted description of the polymerization kinetics at high conversions is not surprising. What is surprising is the behaviour of the initiator efficiency, which is totally different from the expected behaviour based on theoretical considerations (18). As it is shown in Figure 3 there is a large discrepancy between the theoretical values calculated with equation 1a, and those obtained from the data. Furthermore, none of the empirical equations proposed to date, account for the change in efficiency are capable of representing the behaviour shown in Figures 3 through 6 and account for the fate of the initiator fragments (16,18,20,21,23). It is also evident that all of the rate parameters are affected in a similar way by the changes occuring in the reaction environment. The functional relationship that can be used to represent the behaviour shown in Figure 6 is of the form (24):

$$\xi = \exp\left(\frac{\theta X}{XL - X}\right) \quad (39)$$

where X is the conversion and XL is the limiting conversion.

Equation 39 has the structure proposed for the rate constants on the basis of the free volume theory (1,5,9). From this, it would be expected that the models developed from the free volume theory would be very successful in predicting both, the rate behaviour and the molecular properties at high conversions. The reason why these models have been only partially successful stems from the

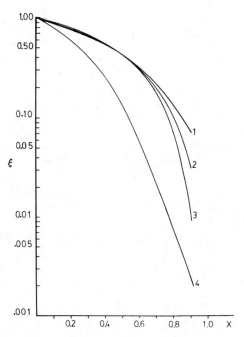

Figure 6. Variation of the rate parameters as functions of conversion: for Case III: (1) $\xi = (k_{tc}/k_p^2)/(k_{tc}/k_p^2)_0$; (2) $\xi = (k_x/k_p)/(k_x/k_p)_0$; (3) $\xi = f/f_0$; (4) $\xi = (k_{tc}/k_p^2)/(k_{tc}/k_p^2)_0$ from reference (18).

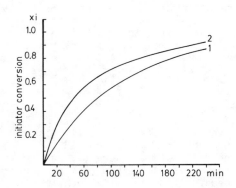

Figure 7. Initiator conversion as function of time: (1) For first order thermal decomposition; (2) Calculated from equation (38).

constraint dictated by equation 24. Preliminary calculations, on the order of magnitude of the terms contributing to the molecular weight averages, indicates that consideration of the radical population in the property equations may relax the constraint on the classical model.

Summary and Conclusions

The instantaneous values for the initiator efficiencies and the rate constants associated with the suspension polymerization of styrene using benzoyl peroxide have been determined from explicit equations based on the instantaneous polymer properties. The explicit equations for the rate parameters have been derived based on accepted reaction schemes and the standard kinetic assumptions (SSH and LCA). The instantaneous polymer properties have been obtained from the cummulative experimental values by proposing empirical models for the instantaneous properties and then fitting them to the cummulative experimental values. This has circumvented some of the problems associated with differenciating experimental data. The results obtained show that:

>None of the models, theoretical or empirical, proposed in the literature to account for the change in efficiency as function of conversion, can describe the behaviour observed.

>The rate parameters follow similar conversion trajectories. Therefore, the rate constants and the initiator efficiency can be modelled with the same equation. An equation of the form of equation 39 is suggested. The theoretical justification for the form of equation 39 stems from the free volume theory.

>From the reaction schemes investigated, it is clear that induced decomposition and primary radical termination reactions should be considered in the initiator balances in order to account for the observed initiator loadings. This is particularly important when relatively high initiator concentrations are involved.

>The use of an overall initiator efficiency, appears to be more effective than the cage effect concept in describing both the effect of the initiator concentration on the efficiency, and the initiator loadings.

From the analysis of the rate equations it can be concluded that the classical polymerization model does not apply whenever the instantaneous polydispersity is greater than 2 or smaller than 3/2. This limitation of the classical model has resulted in piecewise continuous models for high viscosity polymerizations. Preliminary calculations, on the order of magnitude of the terms contributing to the molecular weight averages, indicates that consideration of the radical population in the property equations may relax the constraint on the classical model. We are currently extending our work in this direction and we will report the results in the near future.

Literature Cited
1. Marten, F. L., Hamielec A. E.; "ACS Symposium Series", 104, 43 (1979)
2. Cardenas, J., O'Driscoll, K. F.; J. Polymer Sci. A1, 14, 883 (1976)
3. Cardenas, J., O'Driscoll, K. F.; J. Polymer Sci. A1, 15, 1883 (1977)
4. Cardenas, J., O'Driscoll, K. F.; J. Polymer Sci. A1, 15, 2097 (1977)
5. Harris. B., Hamielec A. E., Marten, F. L.; "ACS Symposium Series", 18, 3199 (1980)
6. Soh, S. K., Sundberg, D. C. ; J. Polymer Sci. Polymer Chem., 20, 1299; 1315; 1331; 1345 (1982)
7. Tulig, T. J., Tirrell, M; Macromolecules, 14, 1501 (1981)
8. Tulig, T. J., Tirrell, M; Macromolecules, 15, 459 (1982)
9. Hamielec, A. E., Friis, N; "Introduction to Chain Polymerization Kinetics", Short Course Notes, Mc Master University (1980)
10. O'Driscoll, K. F., White, P. J; J. Polymer Sci a3, 283 (1985)
11. Berger, K. C., Deb, P. C. and Meyerhoff, G.; Macromolecules, 10 (5), 1075 (1977)
12. Moad, G., Rizzardo, E., Solomon, D. H.; Macromolecules 15, 909 (1982)
13. Bevington, J. C.; Ebdon, J. R.; Huckerby, T. N.; Hutton, N. W. E. Polymer 23, 163 (1982)
14. Garcia-Rubio, L. H.; Ro, N.; Patel, R.; Macromolecules, Vol 17, 1998 (1984)
15. North, A. M., "The Kinetics of Free Radical Polymerization" Pergamon Press (1966)
16. Biesenberger, J. A.; Sebastian, D.H., "Principles of Polymerization Engineering", John Wiley & Sons, New York, (1983)
17. Eastmond, C. G., "Chemical Kinetics, Vol.14a", Bamford, C. H and Tipper, C. F. H editors, Elsevier, (1976)
18. Duerksen, J.H. and Hamiche A.E. J. Polym. Sci. Part C, 155 (1968).
19. Hui, A.T. and Hamielec A.E., J. Polym. Sci. Part C, 167 (1968).
20. Yenalyev, V.D. and Melnichenko, V.I., in "Emulsion Polymerization" eds I. Piirma and J.L. Gardon ACS symp. Series N° 24, 221 (1976).
21. Hui, A. PhD Thesis, McMster University, (1970)
22. Allen, P. E. M; Patrick, C. R., "Kinetics and Mechanisms of Polymerization Reactions", John Wiley & Sons, (1974)
23. Ross, R. T.; Lawrence, L. R., AIChE Symp. Series, 160, Vol 72, 74, (1974)
24. Fabian., E., Private Communication, (1982)

RECEIVED December 2, 1985

# Mathematical Modeling of Emulsion Polymerization Reactors
## A Population Balance Approach and Its Applications

**A. Penlidis, J. F. MacGregor, and A. E. Hamielec**

Department of Chemical Engineering, McMaster Institute for Polymer Production Technology, McMaster University, Hamilton, Ontario, Canada L8S 4L7

Research on the modelling, optimization and control of emulsion polymerization (latex) reactors and processes has been expanding rapidly as the chemistry and physics of these systems become better understood, and as the demand for new and improved latex products increases. The objectives are usually to optimize production rates and/or to control product quality variables such as polymer particle size distribution (PSD), particle morphology, copolymer composition, molecular weights (MW's), long chain branching (LCB), crosslinking frequency and gel content.

Polymer production technology involves a diversity of products produced from even a single monomer. Polymerizations are carried out in a variety of reactor types: batch, semi-batch and continuous flow stirred tank or tubular reactors. However, very few commercial or fundamental polymer or latex properties can be measured on-line. Therefore, if one aims to develop and apply control strategies to achieve desired polymer (or latex) property trajectories under such a variety of conditions, it is important to have a valid mechanistic model capable of predicting at least the major effects of the process variables.

Any attempt to develop control strategies for the previously mentioned product quality variables should consist of four basic stages: (i) specification of reactor/process model forms, (ii) plant experimentation and collection of data, (iii) a stage where the model predictive power is being checked based on the data collected in (ii), and (iv) use of the model's predictive power and extrapolation to other operating conditions, with the final aim to design effective control schemes and devise novel modes of reactor/process operation. In other words, it is very essential to incorporate our knowledge from physics or chemistry (provided, of course, that one has a good understanding of the various phenomena underlying the specific process) into a concise, compact mathematical scheme (the model), which will subsequently be employed to simulate the process, study its special characteristics and predict its performance.

Polymerization processes are not "easy" processes to handle experimentally. There exist a wide variety of operating factors that could cause the production of a latex or polymer with totally different properties than the previous one. Long reaction times and time-consuming analytical techniques to fully characterize a given product make the situation more complicated in that one does not always have the freedom to run a specific experiment or change some conditions in order just to check the process behaviour. In addition, plant personnel are in many cases understandably reluctant to even attempt changes in a specific production system that "seems to work". It is evident, therefore, that if one has at one's disposal a process model which has been tested and evaluated for its predictive powers, then instead of asking specific questions to the real process, one has the advantage to ask these questions to the model. It is needless to say that the latter is a much simpler task, since it can be done with the use of a computer on a simulation of the process. If the answer looks promising, one can then implement the changes in a laboratory or pilot bench-scale reactor first, and if the model prediction is experimentally verified, one can then proceed and apply the changes in the production scale. To summarize, the success of any optimization effort or of any subsequent control effort obviously depends upon having valid dynamic (or, in some cases, steady-state) models of the physical and chemical phenomena occurring in these complex polymerization systems.

The derivation and development of a mathematical model which is as general as possible and incorporates detailed knowledge from phenomena operative in emulsion polymerization reactors, its testing phase and its application to latex reactor design, simulation, optimization and control are the objectives of this paper and will be described in what follows.

Emulsion Polymerization Models

Models for emulsion polymerization reactors vary greatly in their complexity. The level of sophistication needed depends upon the intended use of the model. One could distinguish between two levels of complexity. The first type of model simply involves reactor material and energy balances, and is used to predict the temperature, pressure and monomer concentrations in the reactor. Second level models cannot only predict the above quantities but also polymer properties such as particle size, molecular weight distribution (MWD) and branching frequency. In latex reactor systems, the level one balances are strongly coupled with the particle population balances, thereby making approximate level one models of limited value (1).

In recent years, considerable advances have been made in the modelling of emulsion polymerization reactors. In general, until 1974, models for these systems did not include the particle nucleation phenomena, nor did they consider population balances to account for the PSD's. Now, both homogeneous nucleation (2,3,4), and micellar nucleation mechanisms (collision or diffusion theory) are usually included in the models. Again, two levels of model are used to account for particle size development. The "monodispersed approximation" model is based on modelling the development of the number of polymer particles and the total particle volume. Assuming monodispersed spherical particles, the particle size is calculated as proportional to the cube root of the total volume of polymer phase divided by the number of particles. The surface area of polymer particles (needed to calculate the micelle area) is also obtained as proportional to the two-third power of the volume. Under the monodispersed assumption, it is frequently implicitly assumed that the average number of radicals per polymer particle is constant throughout the reaction and equal to 0.5 (Smith-Ewart kinetics). The second level of emulsion polymerization model employs a population balance approach (or, an age distribution analysis) to obtain the full PSD. By treating the moments of these population balance equations to obtain total or average properties, the set of equations that must be solved is considerably simplified, however, the full set of equations must be solved to determine the PSD.

Two of the most comprehensive discussions of these models were presented by Min and Ray (5) and by Poehlein and Dougherty (6). Min and Ray (5) gave a very general model framework which should be capable of modelling most emulsion polymerization systems. Of course, decisions must be made on the relative importance of the various phenomena occurring in a particular system. Other, more recent efforts on the modelling of emulsion reactors include the ones of Table I. Further details can also be found in (30).

Uses of Models. Future advances in latex production technology will be achieved largely through the efficient use of these comprehensive emulsion polymerization reactor models. Obvious uses of the models include the simulation and design of batch and continuous reactor systems. Novel operating strategies or recipe modifications which give higher productivity or improved quality can be identified. Control policies can be developed through simulation of various schemes of operation, or by using the model directly to develop advanced multivariable control algorithms for a production system in place. Reactor venting systems can be designed, which will reduce the risk of vessel rupture during a runaway polymerization (31). Operators can be trained using iterative graphics terminals along with the simulation model. Analysis of the consequences of system failures or upsets, which is an important safety consideration, can be greatly simplified.

TABLE I. Recent Work on the Modelling of Emulsion Polymerization Reactors

| Literature Source | Polymer System | Level of Model |
|---|---|---|
| Thompson and Stevens (7) | G | PB |
| Cauley et al. (8) | G | PB |
| Kirillov and Ray (9) | MMA | PB |
| Min and Ray (10) | MMA | PB |
| Chiang and Thompson (11) | VAc | ADA |
| Kiparissides et al. (12) | VAc | ADA |
| Min and Gostin (13) | VCM | PB |
| Sundberg (14) | G | PB |
| Kiparissides et al. (15) | VAc | ADA |
| Schork et al. (16) | MMA | MA |
| Ballard et al. (17)* | G | MA |
| Lichti et al. (18) | STY | PB |
| Lin et al. (19)* | STY/AN | MA |
| Hoffman (20)* | SBR | MA |
| Kiparissides and Ponnuswamy (21) | STY | PB |
| Lichti et al. (22) | STY | PB |
| Pollock et al. (23) | VAc | ADA |
| Gilbert and Napper (24) | STY | PB |
| Hamielec et al. (25)* | SBR | PB, MA |
| Lichti et al. (26) | STY | PB |
| Broadhead (27)* | SBR | PB, MA |
| Broadhead et al. (28)* | SBR | PB, MA |
| Gilbert et al. (29) | STY | PB |

* Copolymerization

Symbols: G = general polymer system, PB = population balance model, ADA = age distribution analysis, MA = monodispersed approximation

Finally, a less obvious use of the models is to offer guidance in the selection and development of on- or off-line sensors for reaction monitoring or product characterization, a very interesting aspect which will be briefly discussed later in this paper.

## Model Development

**General.** In this section, a mathematical dynamic model will be developed for emulsion homopolymerization processes. The model derivation will be general enough to easily apply to several Case I monomer systems (e.g. vinyl acetate, vinyl chloride), i.e. to emulsion systems characterized by significant radical desorption rates, and therefore an average number of radicals per particle much less than 1/2, and to a variety of different modes of reactor operation.

The development of the differential equations which describe the evolution of particle size and molecular weight properties during the course of the polymerization is based on the so-called "population balance" approach, a quite general model framework which will be described shortly. Symbols which will be used in the subsections to follow are all defined in the nomenclature.

**Material Balances.** The material (mass) balances for the ingredients of an emulsion recipe are of the general form: (Accumulation) = (Input) - (Output) + (Production) - (Loss), and their development is quite straightforward. Appendix I contains these equations together with the oligomeric radical concentration balance, which is required in deriving an expression for the net polymer particle generation (nucleation) rate, f'(t).

**Population Balance Approach.** The use of mass and energy balances alone to model polymer reactors is inadequate to describe many cases of interest. Examples are suspension and emulsion polymerizations where drop size or particle distribution may be of interest. In such cases, an accounting for the change in number of droplets or particles of a given size range is often required. This is an example of a population balance.

A population balance is a balance on a defined set of countable or identifiable entities in a given system as a result of all phenomena which add or remove entities from the set. If the set in question is the number of droplets between diameter D and (D + dD) in a vessel, the set may receive droplets by flow into the vessel, by coalescence or by growth from smaller droplets. The set may also lose droplets by outflow from the vessel, by coalescence or by growth out of the set's size range.

General discussions of the basic ideas have been presented in (33) and (34). Their applications to crystallization systems were reported in (35) and (36), and to polymerization reactors in (5) and (37) to (41) (see also Table I).

A discussion of the population balance approach in modelling particulate systems and the derivation of the general equation are given in Appendix II.

**Particle Size Development.** Now that a general total property balance equation has been developed (equation (II-9)), one can use it to obtain ordinary differential equations (ode's) which will describe particle size development. What is needed with equation (II-9) is an expression for $dp(t,\tau)/dt$, where p denotes a specific property of the system (e.g. particle size). Such an expression can be written for the rate of change of polymer volume in a particle of a certain class. The analysis, which is general and described in Appendix III, will finally result in a set of ode's for $N_p(t)$, $D_p(t)$, $A_p(t)$ and $V_p(t)$.

**Molecular Weight Development.** The idea of following classes of polymer particles born in the reactor between times $\tau$ and $\tau + d\tau$ can also be employed in the molecular weight development. One can write differential equations for the live (active) and dead polymer chains in the reactor, including contributions from the following reactions: transfer to monomer, transfer to polymer, transfer to a chain transfer agent (CTA) and terminal double-bond polymerization. The result is a series of ode's describing the evolution of the leading moments of the MWD ($Q_0$, $Q_1$ and $Q_2$) and the number of branch points per polymer molecule.

Final Mathematical Model. The final mathematical model thus developed consists of ode's describing the evolution of $x(t)$, $I(t)$, $S_T(t)$, $V_R(t)$ (material balances), of $N_p(t)$, $D_p(t)$, $A_p(t)$, $V_p(t)$ (particle size balances), and of $Q_0(t)$, $Q_1(t)$, $Q_2(t)$ and $B_N(t)$ (molecular weight part). The basic assumptions underlying the model development are summarized below:

(a) Particle agglomeration (coalescence) is neglected.
(b) One polymer particle from a certain class $n(t,\tau) d\tau$ is representative of the whole class and $M_p$ is the same in all classes, due to a very rapid diffusion of monomer from monomer droplets into polymer particles.
(c) The reactor vessel is an ideal CSTR.
(d) During the derivation of an expression for $R_w$, use is made of the collision theory and no distinction between oligomers of different chain lengths is considered. The stationary-state hypothesis for radicals is applicable and the reactivity of a macroradical depends only on the identity of its terminal monomer unit. Effectively, all monomer is consumed by propagation reactions.
(e) Termination reactions are insignificant for the molecular weight development of Case I systems and all rate constants involved are independent of $\tau$, i.e. independent of particle size (58,73).

The next two steps after the development of a mathematical process model and before its implementation to "real life" applications, are to handle the numerical solution of the model's ode's and to estimate some unknown parameters. The computer program which handles the numerical solution of the present model has been written in a very general way. After inputing concentrations, flowrate data and reaction operating conditions, the user has the options to select from a variety of different modes of reactor operation (batch, semi-batch, single continuous, continuous train, CSTR-tube) or reactor start-up conditions (seeded, unseeded, full or half-full of water or emulsion recipe and empty). Then, IMSL subroutine DGEAR handles the numerical integration of the ode's. Parameter estimation of the only two unknown parameters $\varepsilon$ and $D_w$ has been described and is further discussed in (32).

Particle Size Distribution Determination. To consider the full PSD, a population balance or age distribution analysis on particles must be employed. Table II gives a summary of recent work concerning the determination of PSD's in emulsion systems, using both the "monodispersed" approximation and the population balance approach. More details can be found in the literature sources cited in the Table.

The methods suggested in Table II for the calculation of the full PSD require systems of integro-differential, partial differential or difference differential equations to be solved, which in most of the cases is a tedious task requiring considerable computation time. The Min and Ray (5,10) model can give the full PSD and, theoretically, it is the most general mathematical model to date, although their moment method needs a separate set of equations for each additional particle generation, thus making it less convenient for the simulation of a CSTR.

The conceptual approach to modelling the full PSD is to discretize the particle population according to age. Each subdivision is then referred to as "a generation". The diameters of all subsequent particle generations are inferred from the diameter of particles of the first generation and the nucleation time of the "generation" of interest with respect to final time at which the PSD is required. In this way, only the diameter of the particles of the first generation is integrated, as given by (d $d_p(t_f, t_i)$/dt) which was previously derived and where $(t_f, t_i)$ denotes diameter of a particle born at time $t_i$, now at time $t_f$ (27,59).

Then, the PSD is constructed by storing the number of particles and the diameter of a single particle of each generation and is finally presented as a plot of the normalized number frequency of particles versus particle diameter.

Table II. Recent Work on the Modelling of PSD's in Emulsion Polymerization Reactors

| Literature Source | Approach/System |
|---|---|
| Nomura et al. (46) | MONO, VAc |
| Min and Ray (5) | PB, G |
| Dickinson (43) | ADA, STY |
| Poehlein and Dougherty (6) | MONO, G |
| Cauley et al. (8) | PB, G |
| Min and Ray (10) | PB, MMA |
| Kiparissides et al. (12) | ADA, VAc |
| Poehlein (54) | MONO, G |
| Sundberg (14) | PB, G |
| Ray (1) | PB, G |
| Ballard et al. (17) | MONO, G |
| Hoffman (20) | MONO, SBR |
| Kiparissides and Ponnuswamy (21) | ADA, STY |
| Lichti et al. (22) | MONO, PB, STY |
| Nomura and Harada (47) | MONO, VAc |
| Bataille et al. (55) | MONO, STY |
| Lichti et al. (26) | PB, STY |
| Tsai et al. (56) | MONO, STY |
| Broadhead (27) | MONO, ADA, SBR |
| Gilbert et al. (29) | PB, G |
| Hoffman (57) | PB, SAN |
| Pollock (58) | ADA, VAc |

Symbols: MONO = monodispersed approximation, PB = population balance approach, ADA = age distribution analysis, G = general polymer system, VAc = vinyl acetate, STY = styrene, MMA = methyl methacrylate, SBR = styrene butadiene rubber, SAN = styrene acrylonitrile.

The advantage of the PSD calculation method outlined in this subsection is that only one additional differential equation need be integrated, a fact which considerably simplifies matters and saves in computation time. A drawback of the present approach (and also of the other approaches of Table II) is that the statistical variation of the distribution arising from the variation of the number of radicals per particle in a class of a certain size is missing. The PSD's thus calculated indicate the average particle diameter growth of each particle.

Model Applications

General. A mathematical model has been developed in the previous section, which can now be employed to describe the dynamic behaviour of latex reactors and processes and to simulate present industrial and novel modes of reactor operation. The model has been developed in a general way, thus being readily expandable to include additional mechanisms (e.g. redox initiation (59)) or to relax any of the underlying assumptions, if necessary. It is very flexible and can cover various reactor types, modes of operation and comonomer systems. It will be shown in the following that a model is not only a useful mathematical tool but that it can also be very efficiently used in the simulation, optimization, design and control of latex reactors.

Batch and Semi-batch Latex Reactors. The previously developed model has been able to predict the most consistent batch latex reactor data concerning emulsion polymerization of VAc from the literature (46, 48, 60-62). It has also been tested under a wide variety of operating conditions in our 1 gallon stainless steel pilot plant reactors. Results from this experimental investigation, involving predictions of conversion, particle diameter and weight average molecular weight at different temperatures and initiator or emulsifier levels, are reported and discussed in (32).

The model was further employed in an effort to study the nucleation stage 1 in VAc emulsion homopolymerizations and the evolution of the PSD. It is well known that stage 1 has a typical duration of less than 10 minutes, a time interval which is not sufficient to exploit feedback information from the system. Using the model, semi-batch strategies with respect to emulsifier/initiator feedrates were identified, which might theoretically prolong the nucleation stage, cause multi-particle generations and, therefore, multimodal PSD's (59). A combination of a valid mechanistic model like the present one with such optimization techniques as the Pontryagin's maximum principle to theoretically derive open-loop desired flowrate profiles to achieve some prespecified target while minimizing an appropriate objective function which makes physical sense, looks very promising as an immediate future extension, with the final aim to experimentally verify these strategies and, hence, further test the model's validity.

Continuous Reactors and Reactor Trains. The model successfully simulated the dynamic behaviour of continuous latex reactors or trains for the conventional production of poly (VAc) under conditions of sustained property oscillations (58, 63, 64). From the model's applications to continuous "Case I" polymerization systems and from an understanding of the nucleation phenomenon, it became evident (58,59) that the effort to eliminate these industrially undesirable limit cycles should concentrate on the redesign of the reactor train configuration rather than on the strict application of multivariable optimal stochastic control theories to an ill-behaving process. The model was found extremely helpful in testing several reactor configurations (a task which otherwise might have been impossible) and in finally suggesting a new continuous reactor train set-up. The new set-up was theoretically investigated in a simulation study (41, 58, 64), found to eliminate the oscillations and to additionally possess many advantages compared to a conventional train (30). Experimental verification of the redesigned train has been accomplished in (65), where the new system is shown to be free of oscillations and to offer an increased productivity as well as improved final product quality.

On-line State Estimation, Optimal Sensor Selection and Control. Realistically, it is very difficult to have on-line measurements of all the major polymer or latex properties of interest, but perhaps one could rely upon one or two of the sensors available for the on-line measurement of a few states (e.g. conversion). In order to estimate some of the other states (e.g. particle diameter averages), Kalman filters or Observers should be used. A number of papers have investigated these state estimation schemes (58, 66, 67).

The development of on-line sensors is a very costly and time-consuming process. Therefore, if one has available a dynamic model of the reactor which predicts the various polymer (or latex) properties of interest, then this can be used to guide one in the selection and development of sensors. Ideas from the optimal statistical design of experiments together with the present model expressed in the form of a Kalman filter have been successfully used (58) to select those combinations of existing or hypothetical sensors which would maximize the information that could be obtained on the states of the polymerization system. Both the type of sensors and the precisions necessary for them are easily investigated in this way. By changing the choice of the measurement matrix and the measurement covariance matrix, the choice of sensors or sensor combinations which optimize the generalized variance of the state estimates of interest (obtained from the Kalman filter) can be made.

There are much fewer reported applications on the use of mechanistic models for on-line latex reactor control. The problem is not trivial at all and what can be done in this interesting area is the subject of (68).

## Model Extensions

Other Monomer Systems. Very slight modifications are required to make the model applicable to emulsion homopolymerization of vinyl chloride (VCM). An initial study on PVC reactors has been reported in (69) and some more recent results following will finely illustrate the case.

Figure 1 shows conversion-time histories for batch emulsion VCM reactors from (70). The recipes used consisted of 1.0 liter of water, 0.47 liters of VCM and varying amounts of soap and initiator, as indicated on the figure. For the cases of Figure 1b, Berens (70) measured $0.68 \times 10^{18}$ particles per liter of latex for the upper curve (I = 1.0 gr, S = 3.0 gr) and $0.34 \times 10^{17}$ for the lower one corresponding to I = 1.0 gr and S = 1.15 gr. Our model's predictions were $0.2 \times 10^{18}$ and $0.14 \times 10^{17}$, respectively. In Figure 1a, the same amount of emulsifier was used for both runs. Berens (70) estimated $\sim 0.38 \times 10^{17}$ particles per liter of latex for both cases, while our model's prediction was close to $\sim 0.22 \times 10^{17}$.

Liegeois (71), using 400.0 ml of water, 280.0 gr of VCM, 280.0 mgr of potassium persulphate initiator and 2.5 gr of sodium lauryl sulphate emulsifier at 50°C, measured by electron microscopy PVC latex particle diameters close to 190 Å for a conversion level of 24%. The model predicted an average particle diameter equal to 200 Å for the same conversion level and for the same experimental conditions.

In an attempt to further test the predictive powers of our model, three more (continuous) experimental recipes from (70) were simulated. Table III cites the three recipes and Figures 2 and 3 show the obtained results.

Table III. Summary of Seeded CSTR Recipes (70)

| Recipe | A | B | C |
|---|---|---|---|
| Monomer (gr VCM/lit)* | 400 | 400 | 400 |
| Soap (gr/lit) | 1.25 | 2.0 | 2.0 |
| Initiator (gr/lit) | 1.0 | 1.0 | 1.0 |
| Temperature (°C) | 50 | 50 | 50 |
| Residence time (sec) | 3600 | 1800 | 1800 |
| Seed x(t) | 0.05 | 0.045 | 0.1 |
| Seed $N_p(t)$ (#/lit) | $1.0 \times 10^{18}$ | $1.0 \times 10^{18}$ | $2.2 \times 10^{18}$ |
| Seed diameter (Å) | 900 | 300 | 630 |

* In all cases, VCM was fed in excess.

Triangles or inverted triangles in Figures 2 and 3 represent experimentally measured conversion data. The agreement between model predictions and experimental results is very good. Inspecting closer Figures 2 and 3 one can also see the effect of several operating variables, such as soap concentration, residence time and seed characteristics, on the final productivity. As an exercise in this examination of a seeded CSTR, the model prediction at t = 3600 seconds for average particle size was compared to that measured

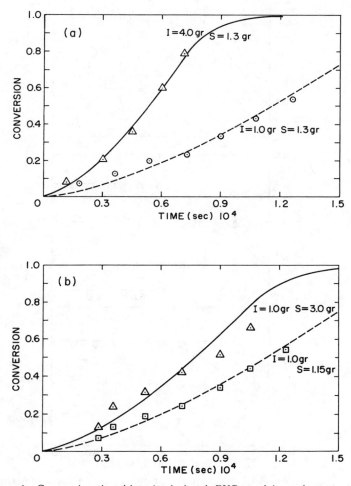

Figure 1. Conversion-time histories in batch PVC emulsion polymerization for different initiator and emulsifier concentrations.

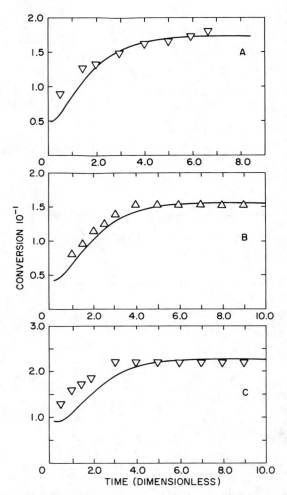

Figure 2. Seeded overflow CSTR starting full of recipe for the production of emulsion PVC (see Table III): Conversion vs. dimensionless time.

for recipe B (70). Berens (70) reported an average diameter close to 1200 Å, whereas the model predicted quite satisfactorily 1070 Å.

A final step during the study of continuous emulsion PVC production with the new split feed reactor system (30, 64, 65) was to again check the predictive validity of our model, testing the following idea: as shown elsewhere (59, 64, 65), the first small reactor in the new train configuration acts as a continuous seeder for the second (i.e. the first large) reactor of the train. If then reactor 1 were operated in such a way so as to produce exactly the same seed characteristics as one of Berens' (70) recipes (say, recipe A), reactor 2 should theoretically give the same results that Berens (70) measured out of his single seeded CSTR, since reactor 2 simply grows the incoming seed particles. The obtained picture is shown in Figure 4. The train produced the same latex as a single seeded CSTR and the model was quite successful in predicting this behaviour. The slight discrepancy at the beginning (see Figure 4), i.e. for short residence times, is most likely due to induction time effects or other start-up experimental details, which were unknown to us from the article.

Copolymer Systems. The copolymer model development followed the homopolymer model development, properly accounting though for the presence of two monomers, two types of radicals and other implications that a comonomer system can give rise to. Information from (72) was found very helpful. Details on the copolymer model development can be found in (59).

To our knowledge, this is the first time that an emulsion copolymerization model has been developed based on a population balance approach. The resulting differential equations are more involved and complex than those of the homopolymer case. Lack of experimental literature data for the specific system VCM/VAc made it impossible to directly check the model's predictive powers, however, successful simulation of extreme cases and reasonable trends obtained in the model's predictions are convincing enough about the validity and usefulness of the mathematical model per se.

Since there was no experimental data available for the copolymer case, the copolymer model was run as a homopolymer one by setting one of the monomer inputs equal to zero. In this way one could test two extreme cases and see if there were any flaws in the model's logic. The results for a batch latex reactor are shown in Figure 5a.

A problem that one encounters when producing a copolymer in a batch reactor is that of copolymer compositional drift. This drift occurs due to the different reactivities of the two monomers. For the VCM/VAc system, $r_1$ (VCM) = 1.68 and $r_2$ (VAc) = 0.23, therefore VCM is the faster reacting monomer. The theoretical curve for the instantaneous copolymer composition, i.e. $F_1$ versus $f_1$, would lie above the 45 degree line for a system of this type, with the drift being in the direction that tends to decrease the mole fraction of the fast monomer. Our model was tested to see whether it could predict the right drift. For the recipes employed, the model prediction is shown in Figure 5b. The portion of the curve generated by the model parallels the theory in that the curve is above the 45 degree line and the drift is in the right direction, as indicated by the arrow.

For a single continuous reactor, the model predicted the expected oscillatory behaviour. The oscillations disappeared when a seeded feed stream was used. Figure 5c shows a single CSTR behaviour when different start-up conditions are applied. The solid line corresponds to the reactor starting up full of water. The expected overshoot, when the reactor starts full of the emulsion recipe, is correctly predicted by the model and furthermore the model numerical predictions (conversion ~ 25%, diameter ~ 1500 Å) are in a reasonable range.

In conclusion, one could say that despite the lack of experimental data with which the copolymer model predictive powers could be tested, its trends when applied to the VCM/VAc system were very reasonable and in agreement with general experience from systems of this type. There are certain aspects of the model (e.g. the expression for the $x_c$ of the system) that should be refined in the near future, but at least the basic structure

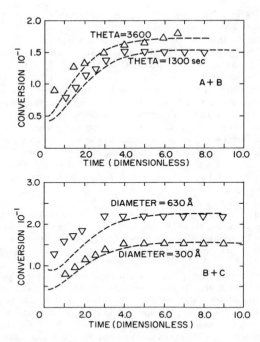

Figure 3. Seeded overflow CSTR, conversions vs. dimensionless time: Effect of process variables (see also Table III).

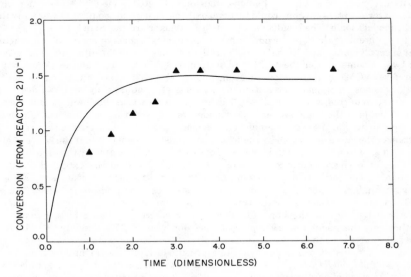

Figure 4. Emulsion PVC production in the new continuous latex reactor train.

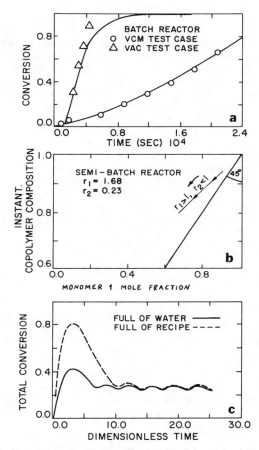

Figure 5. VCM/VAc emulsion copolymerization (a) conversion vs. time in a batch reactor for extreme cases (b) instantaneous copolymer compostion (c) start-up procedures in an unseeded CSTR.

has been successfully set up and could serve as a forerunner for investigating this particular or other copolymer systems.

Conclusions

A valid mechanistic model can be very useful, not only in that it can appreciably add to our process understanding, but also in that it can be successfully employed in many aspects of emulsion polymerization reactor technology, ranging from latex reactor simulation to on-line state estimation and control. A general model framework has been presented and then it was shown how it can be applied in a few of these areas. The model, being very flexible and readily expandable, was further extended to cover several monomer and comonomer systems, in an effort to illustrate some of its capabilities.

Nomenclature

| | |
|---|---|
| $A_m(t)$ | free micellar area |
| $A_p(t)$ | total polymer particle surface area |
| $B_N(t)$ | average number of branch points per polymer molecule |
| CI | concentration of initiator in the initiator stream |
| $d_M, d_p$ | densities of monomer and polymer, respectively |
| $D_p(t)$ | total polymer particle diameter |
| $D_w$ | diffusion coefficient of monomeric radicals in the water phase |
| FI | volumetric flowrate of initiator stream |
| $F_1$ | mole fraction of monomer 1 in the copolymer chain |
| $f_1$ | mole fraction of unreacted monomer 1 in the reactor |
| f(t) | net particle generation rate |
| I(t) | initiator conentration in the reactor |
| k | denotes rate or pseudo-rate constants (subscripts mean: ab = radical capture by particles, d = initiator decomposition, de = radical desorption from particles, h = homogeneous nucleation, m = micellar nucleation, p = propagation, WSI = reaction with water soluble impurities, MSI = reaction with monomer soluble impurities) |
| $k_v$ | ratio of emulsion phase over water phase |
| $M_F(t)$ | concentration of monomer in the feed stream |
| $M_p$ | monomer concentration in the polymer particles |
| MSI(t) | concentration of monomer soluble impurities in the reactor |
| MW | monomer molecular weight |
| $N_A$ | Avogadro's number |
| $N_p(t)$ | total number of polymer particles per liter of latex |
| $n(t,\tau)$ | number of polymer particles in the reactor at time t that were born between times $\tau$ and $\tau + d\tau$ |
| $p(t,\tau)$ | denotes some physical property associated ewith the class of particles $n(t,\tau)$ ($a_p$ = particle area, $d_p$ = particle diameter, $v_p$ = particle volume, v = polymer volume, (t, t) = property at birth, i.e. $t = \tau$). |
| P(t) | total property obtained by summing property $p(t,\tau)$ over all classes of particles in the reactor. |
| $\bar{q}(t,\tau)$ | average number of radicals per particle of the class $(t,\tau)$ |
| $Q_i(t)$ | (i = 0, 1, 2), modified $i^{th}$ moment of the MWD |
| $R_I(t)$ | initiation rate |
| $R_p(t)$ | polymerization rate |
| $R_w^*, R_w^{\bullet}(t)$ | concentration of oligomeric radicals in the reactor |
| $S_{CMC}$ | critical micelle concentration |
| $S_T(t)$ | total soap concentration in the reactor |
| $S_a$ | area coverage by an emulsifier molecule |

| | |
|---|---|
| t | time |
| v(t) | volumetric flowrate |
| $V_p(t)$ | total polymer particle volume |
| $V_R(t)$ | volume of reacting mixture |
| WSI(t) | concentration of water soluble impurities in the reactor |
| x(t) | total monomer conversion level at time t |
| $x_c$ | critical conversion where "stage 2" ends (i.e. separate monomer phase disappears) |
| z | phase coordinates of the phase space |

Greek Letters

| | |
|---|---|
| ε | ratio of $k_{ab}/k_m$ |
| θ | mean reactor residence time |
| ρ(t) | denotes radical production rate (subscripts mean: i = by initiator decomposition, des = by desorption from polymer particles). |
| τ | birth time of polymer particles in the reactor vessel |
| φ(t) | monomer volume fraction in the polymer particles |

Subscripts

| | |
|---|---|
| F | feed conditions |
| IN, in | inlet conditions |
| out | outlet conditions |

## APPENDIX I

## MATERIAL BALANCES

### Initiator Balance

The balance for the number of gmoles of initiator is the following:

$$\frac{dI(t)}{dt} = \frac{CI * FI}{V_R(t)} - I(t)\frac{v_{out}(t)}{V_R(t)} - k_d I(t) - \frac{I(t)}{V_R(t)}\frac{dV_R(t)}{dt} \qquad (I\text{-}1)$$

with

$$\frac{dV_R(t)}{dt} = v_{in}(t) - v_{out}(t) - R_p(t)(MW)\left(\frac{1}{d_M} - \frac{1}{d_p}\right)V_R(t)k_v \qquad (I\text{-}2)$$

Equations (I-1) and (I-2) are true in the general case and can be used to study several modes of reactor operation (e.g. batch, semi-batch, continuous, start-up procedures, etc.). If the assumption is made that the reactor is a vessel continuously operating full, i.e. overflow CSTR, then the right hand side (RHS) of equation (I-2) is zero and (I-1) is considerably simplified to yield:

$$\frac{dI(t)}{dt} = \frac{I_F(t)}{\theta} - \frac{I(t)}{\theta} - k_d I(t) \qquad (I\text{-}3)$$

### Emulsifier Balance

An equation similar to (I-3) can be written for the total concentration of soap (emulsifier) in the reactor, $S_T(t)$, noting of course that the consumption term in the case of soap is zero, since emulsifier is conserved in the reactor.

## Oligomeric Radical Concentration Balance

If $R^\bullet_w(t)$ denotes concentration of (oligomeric) radicals in the reactor, then use of the collision theory to describe radical capture mechanisms, negligible radical termination in the water phase and inflow/outflow of radicals, and application of the radical stationary-state hypothesis would yield the following expression:

$$R^\bullet_w = \frac{\rho(t)}{k_m A_m(t) k_v + k_h + k_{ab} A_p(t) k_v} \qquad (I\text{-}4)$$

where $\rho(t) = (\rho_i(t) + \rho_{des}(t)) N_A$ represents the total radical generation (production) rate.

Water soluble impurities and their effect can be easily included in equation (I-4), through which they are going to directly affect the particle nucleation rate, f(t). If one assumes a first order reaction of an active radical with a water soluble impurity (WSI) to give a stable non-reactive intermediate, then one simply has to add another term in the denominator of equation (I-4), of the form: $k_{WSI} \cdot [WSI](t) \cdot k_v$, and to account for the concentration of WSI with a differential equation as follows:

$$\frac{d[WSI](t)}{dt} = \frac{[WSI]_F(t)}{\theta} - \frac{[WSI](t)}{\theta} - k_{WSI}\,[WSI](t)\,\frac{k_v}{N_A}\,R_W\,. \qquad (I\text{-}5)$$

Since the above treatment would insert an additional parameter in the system (i.e. an additional piece of uncertainty), that of $k_{WSI}$, and since water soluble impurities usually manifest themselves as an induction time for the polymerization reaction, the empirical treatment of (<u>32</u>) might be more than satisfactory.

## Rate of Change of Conversion

The instantaneous conversion x(t) in the reactor can be defined as follows:

$$x(t) = 1 - \frac{M_{MON}(t)}{M_{TOT}(t)} \qquad (I\text{-}6)$$

Combining then a balance for the rate of change of the free (unpolymerized) monomer concentration $M_{MON}(t)$ with one for the total concentration of monomer units $M_{TOT}(t)$ (bounded and unbounded), assuming that the rate of polymerization in the polymer particles is dominant and differentiating equation (I-6), one obtains:

$$\frac{dx(t)}{dt} = \frac{R_p(t)}{M_{TOT}(t)} - \frac{x(t) M_F(t)}{\theta M_{TOT}(t)} \qquad (I\text{-}7)$$

which can eventually be further modified depending upon the reactor start-up conditions.

## APPENDIX II

## POPULATION BALANCES

### Basic Ideas

A phase space is established for a typical particle, whose coordinates specify the location of the particle as well as its quality. Then, ordinary differential equations describe how these phase coordinates evolve in time. In other words, the state of a particle in a processing system is specified by the values of a number of phase coordinates $\underline{z}$. The only requirement on $\underline{z}$ is that they describe the state of the particle fully enough to permit one to write a set of first order ode's of the form:

$$\frac{dz_i}{dt} = u_i(\underline{z}, t); i = 1, 2, \ldots \qquad \text{(II-1)}$$

which show how the $z_i$'s evolve in time while the particle is in the processing system. $u_i$ is the phase velocity and the set of equations (II-1) are the equations of motion for the phase coordinates $z_i$, which may be external (e.g. position in the system) or internal (e.g. size of a particle, number of free radicals in a particle or molecular weight). $z_i$'s may also be continuous or discrete.

A partial differential equation is then developed for the number density of particles in the phase space (analogous to the classical Liouville equation that expresses the conservation of probability in the phase space of a mechanical system) (33). In other words, if the particle states (i.e. points in the particle phase space) are regarded at any moment as a continuum filling a suitable portion of the phase space, flowing with a velocity field specified by the function $u_i$, then one may ask for the density of this fluid streaming through the phase space, i.e. the number density function $n(\underline{z},t)$ of particles in the phase space defined as the number of particles in the system at time t with phase coordinates in the range $\underline{z} \pm (d\underline{z}/2)$.

Then, since in any system accumulation is the net result of both evolution and birth/death processes, and since any latex particle can be characterized by a set of physical quantities which will fully specify a given particle or class of particles, one can obtain the following population balance equation (33):

$$\frac{\partial n(\underline{z}, t)}{\partial t} + \nabla \cdot (\underline{\dot{z}} n(\underline{z}, t)) = f(\underline{z}, t) \qquad \text{(II-2)}$$

where in the general case $\underline{z}$ comprises of external $\underline{x}$ and internal $\underline{r}$ coordinates, i.e. $\underline{z} = [\underline{x}\ \underline{r}]$. The first term on the LHS of equation (II-2) represents accumulation of $n(\underline{z},t)$ and the second term rate of change of $n(\underline{z},t)$ with respect to $\underline{z}$. The RHS function $f(\underline{z},t)$ represents, in the general case, birth and death in a class (i.e. net nucleation) and inflow and outflow in the reactor.

In a well-stirred vessel, the particle environment is uniform and one may accordingly dispense with external coordinates altogether, thus ending up with an one-dimensional phase space and with (II-2) giving:

$$\frac{\partial n(\underline{r}, t)}{\partial t} = \frac{v_0(t) n_0(\underline{r}, t) - v(t) n(\underline{r}, t)}{V_R(t)} - \sum_j \frac{\partial}{\partial r_j}(G_j n(\underline{r}, t) + f(\underline{r}, t)) \qquad \text{(II-3)}$$

where $v_0(t)$ is the inlet volumetric flow rate, $n_0(\underline{r},t)$ the inlet distribution function, $v(t)$ the outlet volumetric flow rate and $n(\underline{r},t)$ is now only a function of the internal phase coordinates $\underline{r}$ and time t. If the tank is an overflow CSTR, then $v_0(t) = v(t)$ and (II-3) becomes simpler.

There is a variety of ways in choosing $\underline{r}$ (5, 40-44). If $\underline{r}$ is set equal to $\tau$, i.e. the birth time of the polymer particles in the reactor vessel, then $n(\underline{r},t)$ becomes $n(t,\tau)$ and $(n(t,\tau)d\tau)$ represents number of particles in the reactor at some time t which were born during the infinitesimal time interval $d\tau$. Integration of $(n(t,\tau)d\tau)$ over the time period t will give the total number of particles in the reactor at time t. Since the particle phase space is now the $\tau$-axis, the analysis becomes an age or residence time distribution analysis, and equation (II-3) simplifies to:

$$\frac{\partial n(t, \tau)}{\partial t} + \frac{\partial n(t, \tau)}{\partial \tau} = f(t, \tau) \qquad \text{(II-4)}$$

with

$$f(t, \tau) = \frac{1}{\theta}\left[n(t, \tau)_F - n(t, \tau)\right] + f_N(t,\tau) + f_{AGG}(t, \tau) \qquad \text{(II-5)}$$

The first term on the RHS of equation (II-5) represents net change by flow, the term $f_N(t,\tau)$ represents contribution by nucleation (i.e. new particle generation) and will be denoted by f(t) from now on for the sake of brevity, and $f_{AGG}(t,\tau)$ is the coalescence (agglomeration) term.

## An Expression for the Nucleation Term f(t)

Deriving an expression for f(t) a considerable simplification occurs if one takes all polymer particles to be nucleated at the same size $d_p(t,t)$. The generation of new polymer particles in an emulsion system is basically due to two mechanisms: micellar and homogeneous particle production. Then, the rate of particle nucleation, f(t), can be expressed as (12):

$$f(t) = k_m A_m(t) R_W \cdot k_v + k_h R_W. \qquad (II-6)$$

In general,

$$A_m(t) = (S_T(t) - S_{CMC}) S_a N_A - A_p(t) + S_p(t) - A_d(t) \qquad (II-7)$$

$S_p(t)$, the area of polymer particles stabilized by polymer end groups rather than soap, might in the general case be important but it is very difficult to obtain an expression for it. $A_d(t)$ on the other hand, the area of monomer droplets, is usually neglected as being quite a few orders of magnitude less than $A_p(t)$.

Expressing the specific homogeneous nucleation rate constant, $k_h$, according to (2) and substituting equation (I-4) into equation (II-6), (II-6) can finally yield a general expression for the particle nucleation rate, f(t), (12,58,59).

## Total Property Balance Equation

Associated with the class of polymer particles $n(t,\tau)d\tau$ in the polymer reactor is a physical property $p(t,\tau)$ (e.g. diameter or area of particles of class $(t,\tau)$, etc.). Then, a total property P(t) (e.g. total particle diameter in the reactor at time t) can be obtained by summing (integrating) $p(t,\tau)$ over all classes of particles in the reactor vessel, viz:

$$P(t) = \int_0^t p(t,\tau) n_{TOT}(t,\tau) d\tau \qquad (II-8)$$

In the general case, $n_{TOT}(t,\tau)$ consists of two types of polymer particles: contribution from newly generated (nucleated) particles which assumed property $p(t,\tau)$ and contribution from newly introduced particles which grew to property $p(t,\tau)$.

Differentiating equation (II-8) with respect to time and using Leibnitz's rule, one can obtain the evolution of P(t) with time. A Laplace transform analysis will finally yield (58,59):

$$\frac{dP(t)}{dt} = \frac{P_{IN}(t)}{\theta} - \frac{P(t)}{\theta} + p(t,t) f(t) + \int_0^t \frac{dp(t,\tau)}{dt} n_{TOT}(t,\tau) d\tau \qquad (II-9)$$

Equation (II-9) simply says that the rate of change of total property P(t) equals the total property inflow minus the total property outflow plus the nucleation term (p(t,t) denotes property at birth, i.e. $t = \tau$) plus the growth terms. Remember that $n_{TOT}(t,\tau)$ denotes, in the general case, particles nucleated in the vessel which grew to the property $p(t,\tau)$ and particles nucleated somewhere else, which entered the reactor with the inlet streams and subsequently grew to $p(t,\tau)$.

## APPENDIX III

## PARTICLE SIZE DEVELOPMENT

<u>Rate of Change of Polymer Volume</u>

The rate of change of polymer volume in a particle of a certain class born at time $\tau$, now being at time t, is given by the following expression:

$$\frac{d\,v(t,\tau)}{dt} = R_p(t,\tau) \frac{(MW)}{d_p} \quad \text{(III-1)}$$

Considering that one particle from a certain class is representative of the whole class and that $M_p$, the concentration of monomer in the polymer particles, is the same for the whole class (for a different approach see (<u>20</u>)), equation (III-1) becomes:

$$\frac{d\,v(t,\tau)}{dt} = \left(\frac{k_p d_M}{N_A d_p}\right) \phi(t) \, \bar{q}(t,\tau) \quad \text{(III-2)}$$

with:

$$\phi(t) = \frac{1-x(t)}{1-x(t)\left(1-\dfrac{d_M}{d_p}\right)}, \quad x(t) = x_c \text{ if } x(t) \leq x_c \quad \text{(III-3)}$$

One can then apply two steady-state balances for radicals, the first for radicals as viewed in the water phase and the second for radicals on a whole class. In other words,

$$\text{Entry = Desorption + Initiation - Water Phase Termination} \quad \text{(III-4a)}$$

$$\begin{pmatrix}\text{Initation of}\\ \text{a class}\end{pmatrix} = \begin{pmatrix}\text{Mutual termination of}\\ \text{radicals in particles}\end{pmatrix} + \begin{pmatrix}\text{Destruction by}\\ \text{impurities}\end{pmatrix} \quad \text{(III-4b)}$$

If one neglects water phase termination, the first of (III-4) becomes:

$$\rho(t,\tau) = \rho_{des}(t,\tau) + R_I(t) \frac{A_n(t,\tau)d\tau}{A_p(t)} \quad \text{(III-5)}$$

with:

$$\rho_{des}(t,\tau) = k_{de}(t,\tau) \, \bar{q}(t,\tau) \, n(t,\tau)d\tau \quad \text{(II-6)}$$

and

$$A_n(t,\tau)d\tau = a_p(t,\tau) \, n(t,\tau)d\tau \quad \text{(II-7)}$$

The second of (III-4) can be written as:

$$R_I(t) \frac{A_n(t,\tau)d\tau}{A_p(t)} = 2\left[\rho(t,\tau) - \rho_{IM}(t,\tau)\right] q(t,\tau) + \rho_{IM}(t,\tau) \quad \text{(III-8)}$$

with

$$\rho_{IM}(t,\tau) = k_{MSI}[MSI](t) \, \bar{q}(t,\tau) \, n(t,\tau)d\tau \quad \text{(III-9)}$$

The first term on the RHS of (III-8) represents mutual termination of radicals in the polymer particles (i.e. second order termination). The second term represents a first order termination of radicals in the polymer particles by monomer soluble impurities (MSI), which are present in the polymer particles due to their transfer in there with monomer during the monomer diffusion phase from monomer droplets.

Combination of (III-5) and (III-8) will yield an expression for $\bar{q}(t,\tau)$. The desorption rate constant, $k_{de}(t,\tau)$, which is present in $\rho_{des}(t,\tau)$, equation (III-6), can be expressed according to (45), (46) and (47). The gel-effect can also be included in the general expression, following (48) to (53).

Rate Expression for Particle Volume

Recalling that

$$v_p(t, \tau) = \frac{v(t, \tau)}{1 - \phi(t)} \qquad (III-10)$$

the rate of change of particle volume is now obtained as:

$$\frac{dv_p(t, \tau)}{dt} = \frac{1}{1 - \phi(t)} \frac{dv(t,\tau)}{dt} + \frac{v(t, \tau)}{(1 - \phi(t))^2} \frac{d\phi(t)}{dt} \qquad (III-11)$$

The LHS term of equation (III-11) represents the term $(dp(t,\tau)/dt)$ in equation (II-9) and consists of a growth term (first term on the RHS) and a shrinkage term (second term of the RHS of (III-11)).

Therefore, the general property balance equation (II-9) becomes for total particle volume:

$$\frac{dV_p(t, \tau)}{dt} = \frac{V_{pIN}(t)}{\theta} - \frac{V_p(t)}{\theta} + v_p(t, t) f(t) + \int_0^t \frac{dv_p(t, \tau)}{dt} n(t, \tau) d\tau \qquad (III-12)$$

Recalling that $v_p(t,\tau) = (1/6) \pi d_p^3(t,\tau)$ and that $a_p(t,\tau) = \pi d_p^2(t,\tau)$, one can derive similar equations for total particle diameter and surface area. The equation for the total number of particles is very straightforward and is based on (II-9) without the growth term.

Literature Cited

1. Ray, W.H. IFAC PRP-4, Ghent, Belgium, 1980, p. 587.
2. Fitch, R.M.; Tsai, C.H. In "Polymer Colloids"; Fitch, R.M., Ed.; Plenum Press: New York; 1971; p. 103.
3. Hansen, F.K.; Ugelstad, J. J. Poly. Sci. 1978, 16, 1953.
4. Hansen, F.K.; Ugelstad, J. In "Emulsion Polymerization"; Piirma, I., Ed.; Academic Press: New York; 1982; p. 51.
5. Min, K.W.; Ray, W.H. J. Macro. Sci.-Revs. Macro. Chem. 1974, C11 (2), 177.
6. Poehlein, G.W.; Dougherty, D.J. Rubber Chem. and Techn. 1977, 50, 601.
7. Thompson, R.W.; Stevens, J.D. Chem. Eng. Sci. 1977, 32, 311.
8. Cauley, D.A.; Giglio, A.J.; Thompson, R.W. Chem. Eng. Sci. 1978, 33, 979.
9. Kirillov,V.A; Ray, W.H. Chem. Eng. Sci. 1978, 33, 1499.
10. Min, K.W.; Ray, W.H. J. Appl. Poly. Sci. 1978, 22, 89.
11. Chiang, A.S.T.; Thompson, R.W. AIChE J. 1979, 25, 552.
12. Kiparissides, C.; MacGregor, J.F.; Hamielec, A.E. J. Appl. Poly. Sci. 1979, 23, 401.
13. Min, K.W.; Gostin, H.I.; Ind. Eng. Chem. Prod. Des. Dev. 1979, 18 (4), 272.
14. Sundberg, D.C. J. Appl. Poly. Sci. 1979, 23, 2197.

15. Kiparissides, C.; MacGregor, J.F.; Hamielec, A.E. CJChE 1980, 58, 48.
16. Schork, F.J.; Chu, G.; Ray, W.H. 1980 AIChE Meeting; Chicago, IL; 1980.
17. Ballard, M.J.; Napper, D.H.; Gilbert, R.G. J. Poly. Sci. 1981, 19, 939.
18. Lichti, G.; Hawkett, B.S.; Gilbert, R.G.; Napper, D.H.; Sangster, D.F. J. Poly. Sci. 1981, 19, 925.
19. Lin, C.C.; Ku, H.C.; Chiu, W.Y. J. Appl. Poly. Sci. 1981, 26, 132.
20. Hoffman, T.W. Annual Polymer Course Notes; Ch. 4; McMaster University; 1981.
21. Kiparissides, C.; Ponnuswamy, S.R. Chem. Eng. Commun. 1981, 10, 283.
22. Lichti, G.; Gilbert, R.G.; Napper, D.H. J. Poly. Sci. 1983, 21, 269.
23. Pollock, M.J.; MacGregor, J.F.;Hamielec, A.E. In "Computer Applications in Applied Polymer Science"; Provder, T., Ed.; ACS Symp. Ser. 1981; Vol. 197; p. 209.
24. Gilbert, R.G.; Napper, D.H. JMS - Rev. Macromol. Chem. Phys. 1983, C23 (1), 127.
25. Hamielec, A.E.; MacGregor, J.F.; Broadhead, T.O.; Kanetakis, J.; Wong, F.Y.C. ACS Rubber Division 123rd Toronto Meeting; May 1983.
26. Lichti, G.; Gilbert, R.G.; Napper, D.H. J. Poly. Sci.: Poly. Chem. Ed. 1980, 18,297.
27. Broadhead, T.O. M.Eng. Thesis, McMaster University, Hamilton, Ontario, 1984.
28. Broadhead, T.O.; Hamielec, A.E., MacGregor, J.F. Macromol. Chem. 1985; accepted for publication.
29. Gilbert, R.G.; Feeney, P.J.; Napper, D.H. 57th CAN-AM CIC Conference, June 1984, Montreal, Canada.
30. Penlidis, A.; MacGregor, J.F.; Hamielec, A.E. AIChE J. 1985, 31 (6), 881.
31. J.E. Huff, Plant/Operations Progress 1982, 1 (4), 211.
32. Penlidis, A.; MacGregor, J.F.; Hamielec, A.E. Polymer Proc. Eng. 1985, 3 (3).
33. Hulburt, H.M; Katz, S. Chem. Eng. Sci. 1964, 19, 555.
34. Randolph,A.D. CJChE 1964, 42, 280
35. Moyers, C.G.; Randolph, A.D. AIChE J. 1973, 19, 1089.
36. Chang, R.Y.; Wang, M.L. Ind. Eng. Chem. Proc. Des. Dev. 1984, 23 (3), 463.
37. Stevens, J.D.; Funderburk, J.O. Ind. Eng. Chem. Proc. Des. Dev. 1972, 11, 360.
38. Thompson, R.W.; Stevens, J.D. Chem. Eng. Sci. 1977, 32, 311.
39. Cauley, D.A.; Giglio, A.J.; Thompson, R.W. Chem. Eng. Sci. 1978, 33, 979.
40. Kiparissides, C.; MacGregor, J.F. Hamielec, A.E. J. Appl. Poly. Sci. 1979, 23, 401.
41. Pollock, M.J.; MacGregor, J.F.; Hamielec, A.E. In "Computer Applications in Applied Polymer Science", Provder, T., Ed.; ACS Symp. Ser. 1981; Vol. 197; p. 209.
42. Gorber, D.M. Ph.D. Thesis, University of Waterloo, Waterloo, Ontario, 1973.
43. Dickinson, R.F. Ph.D. Thesis, University of Waterloo, Waterloo, Ontario, 1976.
44. Dickinson, R.F.; Gall, C.E. 59th CIC Conference, London, Ontario, Canada, 1976.
45. Harada, M.; Nomura, M.; Eguchi, W.; Nagata, S. J. Chem. Eng. Japan 1971, 4 (1), 54.
46. Nomura, M.; Harada, M.; Nakagawara, K.; Eguchi, W.; Nagata, S. J. Chem. Eng. Japan 1971, 4 (2), 160.
47. Nomura, M.; Harada, M. J. Appl. Poly. Sci. 1981, 26, 17.
48. Friis, N.; Nyhagen, L. J. Appl. Poly. Sci. 1973, 17, 2311.
49. Friis, N.; Hamielec, A.E. In "Emulsion Polymerization"; Piirma, I.; Gardon, J.L., Eds.; ACS Symp. Ser. 1976; Vol. 24; p. 82.
50. Yasuda, H.; Lamaz, C.E.; Peterlin, A. J. Poly. Sci. 1971, A2 (9), 1117.
51. Zollars, R.L. In "Emulsion Polymerization of Vinyl Acetate"; El-Aasser, M.S., Vanderhoff, J.W., Eds.; Applied Science Publishers Ltd., 1981; p. 31.
52. Litt, M.H.; Chang, K.H.S. Symp. on Emulsion Polymerization of Vinyl Acetate, Lehigh University, 1980.
53. Friis, N.; Hamielec, A.E. Annual Polymer Course Notes, McMaster University, 1977; Ch. 5; Part I.
54. Poehlein, G.W. In "Polymerization Reactors and Processes"; Henderson, I.N.; Bouton, T.C. Eds.; ACS Symp. Ser. 1979; Vol. 104; p. 1.

55. Bataille, P.; Van, B.T., Pham, Q.B. J. Poly. Sci.: Poly. Chem. Ed. 1982, 20, 795.
56. Tsai, M.C.; Chiu, W.Y.; Lin, C.C. "PACHIEC"; 1983; Vol. II; 76.
57. Hoffman,E.J. M.Eng. Thesis, McMaster University, Hamilton, Ontario, 1984.
58. Pollock, M.J. Ph.D. Thesis, McMaster University, Hamilton, Ontario, 1984.
59. Penlidis, A. Ph.D. Thesis, McMaster University, Hamilton, Ontario, 1985.
60. Keung, C.K.J. M.Eng. Thesis, McMaster University, Hamilton, Ontario, 1974.
61. Singh, S.; Hamielec, A.E. J. Appl. Poly. Sci. 1978, 22, 577.
62. Friis, N.; Hamielec, A.E. J. Appl. Poly. Sci. 1975, 19, 97.
63. Kiparissides, C.; MacGregor, J.F.; Hamielec, A.E. CJChE 1980, 58, 48.
64. MacGregor, J.F.; Hamielec, A.E.; Penlidis, A.; Pollock, M.J. IFAC PRP-5, Antwerp, Belgium, 1983, p. 291.
65. Penlidis, A.; MacGregor, J.F.; Hamielec, A.E., unpublished data; some data presented at the 189th ACS Meeting, May 1985, Miami Beach.
66. Schuler,H. IFAC PRP-4, Ghent, Belgium, 1980, p. 369.
67. Kiparissides, C.; MacGregor, J.F.; Hamielec, A.E. AIChE J. 1981, 27 (1), 13.
68. MacGregor, M.F.; Penlidis, A.; Hamielec, A.E. Poly. Proc. Eng. 1984, 2, 179.
69. Penlidis, A.; Hamielec, A.E.;MacGregor, J.F. J. Vinyl Techn. 1984, 6 (4), 134.
70. Berens,A.R. J. Appl. Poly. Sci. 1974, 18, 2379.
71. Liegeois, J.M. Die Ange. Makromol. Chemie 1976, 56, 115.
72. Hamielec, A.E.; MacGregor, J.F. Proc. Int. Berlin Workshop on Polymer Reaction Engineering, Berlin, West Germany, 1983.
73. Hamielec, A.E. In "Emulsion Polymerization of Vinyl Acetate"; El-Aasser, M.S.; Vanderhoff, J.W., Eds.; Applied Science Publishers Ltd.; 1981; p. 49.

RECEIVED April 18, 1986

# Kinetics Analysis of Consecutive Reactions Using Nelder–Mead Simplex Optimization

Gary M. Carlson[1] and Theodore Provder[2]

Glidden Coatings and Resins, SCM Corporation, Strongsville, OH 44136

>   The Nelder-Mead simplex minimization routine is used to
>   determine kinetics constants for a reaction system
>   involving a first order generation of reactive
>   intermediate B followed by a second order consumption of
>   B by coreaction with species C. Experimentally, the
>   concentration of B is monitored as a function of time.
>   Theoretical curves describing the concentration of B are
>   generated using an Euler-Romberg solution of the
>   differential equations. The Nelder-Mead simplex
>   minimization routine optimizes the values of $k_1$ and $k_2$
>   to obtain the best agreement between theoretical curves
>   and actual experimental data. The application of this
>   technique to the isothermal deblocking and cure of
>   blocked isocyanate containing materials monitored by
>   Fourier transform infrared spectroscopy is demonstrated.

Crosslinking of many polymers occurs through a complex combination of consecutive and parallel reactions. For those cases in which the chemistry is well understood it is possible to define the general reaction scheme and thus derive the appropriate differential equations describing the cure kinetics. Analytical solutions have been found for some of these systems of differential equations permitting accurate experimental determination of the individual rate constants.

Often analytical solutions do not exist, or have not been determined. It is sometimes possible, in these cases, to arrive at approximate expressions or to define conditions such that rate constants or functions of the rate constants can be determined experimentally.

[1] Current address: Research and Development, Ashland Chemical, Columbus, OH 43216
[2] Author to whom correspondence should be addressed

The use of blocked isocyanates to cure hydroxyl containing coatings is an example of a complex system having many practical applications. The chemistry of blocked isocyanates has been reviewed previously by Wicks (1,2). The cure reaction proceeds via consecutive first and second order reactions:

$$\text{R-NCO-B} \xrightarrow{k_1} \text{R-NCO} + \text{BH} \quad (1)$$

$$\text{R-NCO} + \text{ROH} \xrightarrow{k_2} \text{R-NCOOR} \quad (2)$$

Deblocking liberates free isocyanate which reacts with hydroxyl functionality in a second order reaction to form the actual crosslinks. Anagnastou and Jaul have studied the deblocking of MDI derivatives using differential scanning calorimetry (3). The deblocking of coatings systems containing the blocked isocyanates has been investigated previously by Kordomenos and coworkers (4) using thermogravimetric analysis.

Previous investigations (5-8) using Fourier transform infrared analysis to determine the chemical changes which occur during linear heating supported the proposed scheme and showed that, in the absence of catalysts, substantial amounts of free isocyanate were generated. Other investigators have also observed similar behavior in similar coatings systems (9, 10).

The effective development of blocked isocyanate based coatings requires a complete understanding of the cure chemistry. Many materials have been identified which will improve cure response. Often these are reported as deblocking catalysts even though no direct evidence exists to support this claim. Most of these materials are well known catalysts of the cure reaction between hydroxyl and isocyanate and may be improving cure response solely by catalyzing this reaction. Effective development of catalyst systems requires a better understanding of the effect of catalysts on cure.

Modification of the FT-IR analysis techniques to analyze coatings under isothermal cure conditions provides the data needed to determine the rate constants for each reaction. An effective method to generate the rate constants from the experimental data has been found and will be described.

Using the previously proposed chemical scheme (Equation 1-2) and assuming that there is only one isocyanate species, one hydroxyl species, and that the deblocking is irreversible (i.e., the blocking agent leaves the film immediately) then the concentration of each species as a function of time is given by the following differential equations:

$$-d[\text{R-NCO(B)}]/dt = k_1[\text{R-NCO(B)}] \quad (3)$$

$$-d[\text{R-NCO}]/dt = K_2[\text{R-NCO}][\text{OH}] - k_1[\text{R-NCO(B)}] \quad (4)$$

$$-d[\text{OH}]/dt = k_2[\text{R-NCO}][\text{OH}] \quad (5)$$

$$d[\text{R-NCOOR}]/dt = k_2[\text{R-NCO}][\text{OH}] \quad (6)$$

In the absence of simplifying assumptions a non-closed form analytical solution of these equations giving the concentration of each species as a function of time has been accomplished through the use of Bessel functions (11). Simplifying assumptions, such as a large hydroxyl excess, are acceptable theoretically but are not helpful in determining the rate constants of coatings materials with standard compositions.

As seen in Equation 4, the concentration of free isocyanate as a function of time is dependent on both rate constants. Measuring the concentration of the isocyanate as a function of time during isothermal cure provides a convenient method to determine both rate constants.

Many practical applications of cure characterization involve samples for which the data required to convert isocyanate absorbance to concentration is unavailable. The emphasis is often placed on rapid analysis of many samples rather than an exhaustive characterization of a single sample. It is particularly desirable to develop a procedure which can determine the rate constants describing the cure reaction without converting the infrared absorbance curve to concentration. This has been accomplished by normalizing the data in such a way that the rate constants are determined from the shape of the cure curve.

The experimental technique used to obtain the cure curves and a detailed examination of the procedure used to generate the rate constants, an iterative procedure to determine the best values of $k_1$ and $k_2$ based on the set of differential equations describing the cure process, for these systems will be discussed.

## Experimental

Spectra were obtained using a Digilab FTS-15E Fourier Transform Spectrophotometer. A NaCl crystal mounted in a heated cell (Model #018-5322 Foxboro/Analabs, N. Haven, Ct.) was placed in the infrared beam and the chamber allowed to purge for several minutes while the cell was brought to the desired temperature. The temperature of the cell was controlled using a DuPont 900 Differential Thermal Analyzer interfaced to the spectrometer cell. A chlorobenzene solution (ca. 10% by wt.) of the sample was then applied to the crystal using cotton tipped wood splint.

Immediately after applying the sample, spectral data acquisition was started. The small size of the sample, in relation to the heated cell assembly, assures nearly instantaneous equilibration at the desired temperature.

Twenty scans were co-added to produce one spectrum at a resolution of 4 $cm^{-1}$ every 30 seconds. The reaction was followed by monitoring the absorbance of isocyanate as a function of time. Film thickness changes were compensated for by normalizing the isocyanate absorbance to the 1446 $cm^{-1}$ band which remains constant in absorbance during the reaction. The infrared absorbance of the free isocyanate is converted to concentration by comparison to the absorbance observed in a similar sample which has no functional groups with which the isocyanate can react.

## Results and Discussion

The concentration of each chemical species, as a function of time, during cure can be calculated numerically from Equations 3-6 using the Euler-Romberg integration method if the initial concentrations of blocked isocyanate and hydroxyl functionality are known. It is a self-starting technique and is generally well behaved under a wide variety of conditions. Details of this numerical procedure are given by McCalla (12).

In the case under study, the rate constants are unknown but a trial and error approach can be used to determine the rate constants which give the best agreement to the experimental data. The Nelder-Mead simplex algorithm is used to choose the trial and error values (13). This algorithm has been shown to be very useful in other related experiments (14). The procedure starts using an initial guess for the individual rate constants $k_1$ and $k_2$. The Euler-Romberg integration method is then used to generate concentration curves based on the trial rate constants. An objective function describing the difference between the experimental and trial curves is calculated and returned to the Nelder-Mead simplex routine. This objective function is then used, in combination with the other previous best guesses to arrive at a new pair of rate constants. The process is repeated until the values of $k_1$ and $k_2$ which best describe the experimental data have been identified. The complete process is illustrated in Figure 1.

The simplest case for determining rate constants occurs when the starting concentrations of blocked isocyanate and hydroxyl functionality are accurately known. In this case the objective function F, is defined as

$$F = \sum_{t=t_o}^{t=t_f} ( [NCO]_e - [NCO]_c )^2 \qquad (7)$$

where $t_o$ is the initial time, $t_f$ is the final time, $[NCO]_e$ is the experimental isocyanate concentration at time t and $[NCO]_c$ is the calculated concentration at time t.

The effectiveness of the above described computational procedure was tested by generating an analytical ("ideal") data curve by calculating the isocyanate concentration as a function of time assuming rate constants of $k_1$ = 1.0/min and $k_2$ = 1.0 ℓ/mol/min and initial concentrations of blocked isocyanate and hydroxyl of 1.0 M. The objective function, for various values of $k_1$ and $k_2$, for this "ideal" data was calculated and a contour plot for constant values of F was generated and is shown in Figure 2.

This plot is a map of the surface over which the search for the best values occurs. The solution is located in the objective function 'well' at $k_1$ = 1.0/min and $k_2$ = 1.0 ℓ/mol/min. The path taken by the program when the initial guesses of $k_1$ = 1.5/min and $k_2$ = 1.5 ℓ/mol/min also is shown in Figure 2. This starting point is located on a slight slope and begins its descent towards the minimum value with only one major deviation, occcurring after the eighth iteration. The contour plot also demonstrates that no local minima occur over the region in which reasonable answers occur.

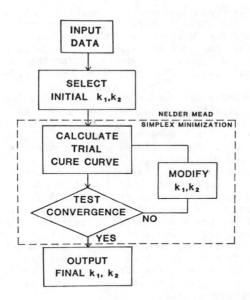

FIGURE 1    KINETICS ANALYSIS FLOW CHART

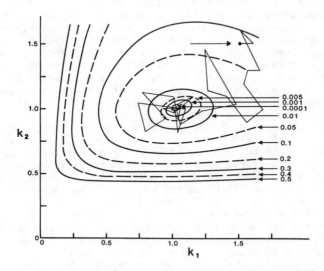

FIGURE 2    CONTOUR PLOT AND SEARCH TRAJECTORY FOR TEST DATA
(INITIAL ESTIMATE: $k_1$ = 1.5 /min, $k_2$ = 1.5 ℓ/mol/min)

The path taken when the search is started at $k_1 = 0.5$/min and $k_2 = 0.5$ ℓ/mol/min is shown in Figure 3. This search begins on a steeper slope and the function, in this case, spirals in on the minimum value. Again the correct values of $k_1 = 1.0$/min and $k_2 = 1.0$ ℓ/mol/min are found. Total computer time for this analysis averaged 3-5 minutes with a Digital Equipment Corporation PDP-11/44 minicomputer.

An experimentally determined curve describing the concentration of free isocyanate during isothermal curing of a blocked isocyanate containing coating is shown in Figure 4. From the minimization procedure the rate constants were determined to be $k_1 = 1.86$ /min and $k_2 = 0.35$ ℓ/mol/min. The predicted concentration is shown by the solid line labeled by squares in Figure 4. It is also possible knowing the rate constants to calculate the concentration of blocked isocyanate and the concentration of crosslinks as a function of time.

Interestingly, it is seen in Figure 4 that the maximum concentration of free isocyanate observed during the deblocking corresponds to 69.2 percent of the total isocyanate. The blocked isocyanate has decreased to a negligible level after only 3 minutes but at the time of complete deblocking only about one-half of the total potential crosslinks have formed. In a coating such as this a better catalyst for deblocking will have little effect on the overall cure rate of the coating. A more effective choice for catalyst would be a catalyst for the cure reaction.

The contour plot for this experiment is shown in Figure 5. The shape of the surface is somewhat different than for the 'ideal' data. This is partly a result of the experimental error which is present in the data and raises the value of the objective function for any trial set of rate constants. There also is a difference in the shape of the surface. The contours for the 'ideal' data, shown in Figures 2 and 3, described a deep circular well with nearly concentric circles surrounding it. The surface of the data is more elongated in the $k_1$ axis. Thus a wide range of $k_1$ values will permit the answer to fall within a given objective function value. The rate constant for cure has a smaller range of acceptable answers. This gives an estimate of the margin of error for each determined rate constant. In this data, $k_2$ is known, to a better degree than $k_1$.

The objective function, F, as a function of iteration, during the search is shown in Figure 6. The objective function decreases rapidly and has neared the minimum value after approximately 20 iterations. The trial values for the rate constants are shown in Figure 7. The values for $k_2$ level off after about 30 iterations, while $k_1$, which has less effect on the value of the objective function, does not level off until about 60 iterations. The convergence criteria is not met until almost 120 iterations. Since the rate constants are essentially determined after 60 iterations, the computer time could be decreased without affecting the accuracy of the final answer by using a less stringent criteria.

The residual plot of the difference between the best computed curve and the experimental data is shown in Figure 8. The largest variance is observed in the vicinity of the maximum isocyanate absorbance and probably arises because of the large changes in concentration which arise during the early stages of cure.

Using the previously described procedure, determination of the rate constants can be made when an accurate curve describing the

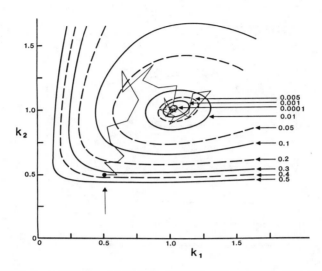

FIGURE 3    CONTOUR PLOT AND SEARCH TRAJECTORY FOR TEST DATA
(INITIAL ESTIMATE: $k_1$ = 0.5 /min, $k_2$ = 0.5 ℓ/mol/min)

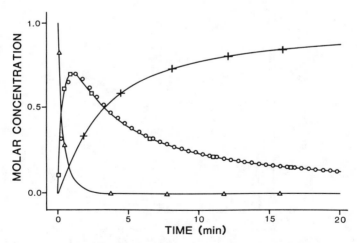

FIGURE 4    EXPERIMENTAL DATA AND CALCULATED ISOTHERMAL CURE
CURVES (△ PREDICTEDD [BLOCKED ISOCYANATE], O EXPERIMENTAL [NCO],
□PREDICTED [NCO], + PREDICTED [CROSSLINKS])

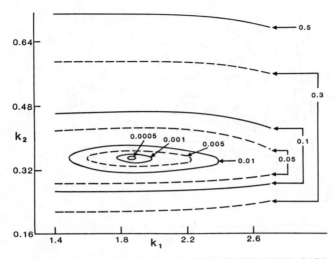

FIGURE 5   CONTOUR PLOTS FOR EXPERIMENTAL DATA

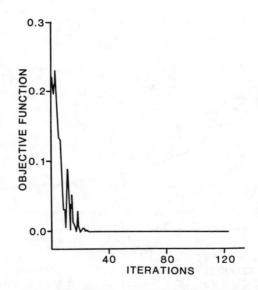

FIGURE 6   OBJECTIVE FUNCTION TRAJECTORY DURING MINIMIZATION

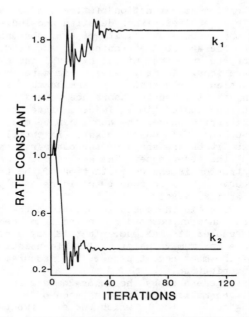

FIGURE 7   RATE CONSTANT TRAJECTORY DURING MINIMIZATION

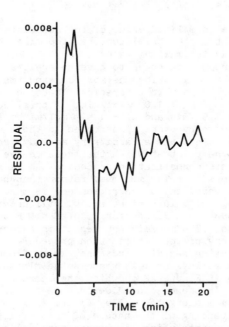

FIGURE 8   RESIDUAL FOR FITTED CURVE

isocyanate concentration as a function of time can be obtained experimentally. This requires either detailed knowledge of the resin system, including the molar absorbtivity of the isocyanate group, required to convert the absorbance readings to actual concentration. It is also necessary to know the initial hydroxyl and blocked isocyanate concentrations. In many cases, this data is unavailable.

It would be particularly useful to determine the rate constants for each step using the isocyanate absorbance versus time rather than concentration as a function of time. Indeed, since the absorbance is essentially an arbitrary measure, the curve can be measured in any convenient units such as FT-IR chart height. Essentially, the procedure bases its fit on the shape of the curve rather than on the actual magnitude of the absorbance. The key to using absorbance rather than concentration is the normalization of the trial curve and the experimental curve. This is done by dividing each point of the trial curve by an arbitrary value.

This arbitrary value is the measured height on the trial curve of the isocyanate absorbance corresponding to the time when the experimental data reaches its maximum. In this way a curve is obtained which has a maximum value of 1.0 in dimensionless units. Since the units of the measurement cancel, the measurement can be done using any convenient units.

Each trial curve generated by the Euler-Romberg integration method is similarly normalized by dividing each of its points by a value corresponding to the time at which the experimental curve reached its maximum.

$$f(t)_{new} = f(t)_{old} / f(tmax) \tag{8}$$

where $f(t)_{new}$ is the normalized value of the calculated curve and $f(tmax)$ is the height of the trial curve at the time when the experimental data had its maximum value.

This process forces each curve to have an arbitrary concentration value of 1.0 at the time in which the maximum experimental absorbance was measured. Figure 9 shows a theoretical 'data' curve for the case where $k_1$ = 1.0 /min and $k_2$ = 1.0 ℓ/mol/min. A trial curve also is depicted where $k_1$ = 1.5 /min and $k_2$ = 1.5 ℓ/mol/min. After normalization the two curves are transformed to those shown in Figure 10. The 'trial' curve now has an arbitrary maximum greater than 1.0 but passes through the point at which the 'data' curve was maximum.

The effect of this normalization procedure can be seen in the contour plot of Figure 11. The minimum, rather than being a 'well' as in the procedure based on concentration now is more of a 'valley' in which a wide range of values of $k_1$ and $k_2$ will provide reasonable solutions to the equation. Values for $k_1$ of from .8 to 1.3 /min and for $k_2$ of from .5 to 1.5 ℓ/mol/min can result in answers with F = 0.005. The trajectory of the minimization procedure is shown in Figure 11. The function rapidly finds the valley floor and then travels through the valley until it reaches the minimum. A similar trajectory is shown in Figure 12 in which the search is started from a different point. In the case of "ideal" data the procedure will still find the minimum along the valley floor.

A better understanding of the nature of this 'valley' is obtained by looking at other points which lie within the 'valley'. The rate constants $k_1$ = 1.25 /min and $k_2$ = 0.7 ℓ/mol/min also lie within the

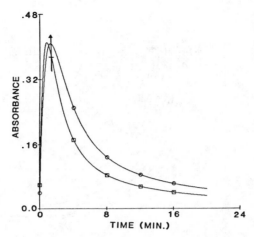

FIGURE 9    TYPICAL TRIAL CURVE BEFORE NORMALIZATION (□ $k_1$ = 1.5/min, $k_2$ = 1.5 ℓ/mol/min TRIAL CURVE; O $k_1$ = 1.0/min, $k_2$ = 1.0 ℓ/mol/min DATA CURVE)

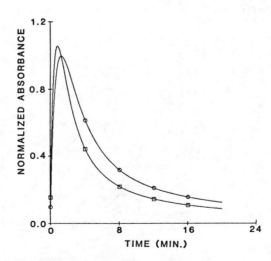

FIGURE 10    NORMALIZED TRIAL CURVE (□ $k_1$ = 1.5/min, $k_2$ = 1.5 ℓ/mol/min TRIAL CURVE; O $k_2$ = 1.0/min, $k_2$ = 1.0 ℓ/mol/min DATA CURVE)

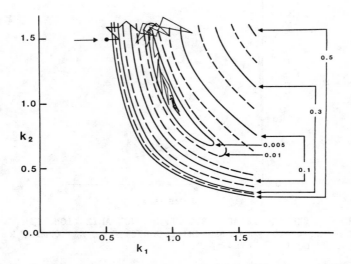

FIGURE 11  CONTOUR PLOT AND SEARCH TRAJECTORY USING NORMALIZATION (INITIAL ESTIMATE: $k_1$ = 0.5 /min, $k_2$ = 1.5 ℓ/mol/min)

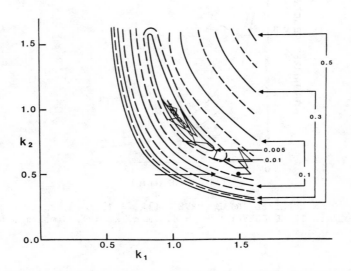

FIGURE 12  CONTOUR PLOT AND SEARCH TRAJECTORY USING NORMALIZATION (INITIAL ESTIMATE: $k_1$ = 1.5 /min, $k_2$ = 0.5 ℓ/mol/min)

valley where the objective function F <0.005. The concentration of isocyanate as a function of time for this case is shown in Figure 13 along with the data for the best fit. It is seen that both attain their maximum isocyanate functionality at nearly the same time. As the normalization procedure relies on the time at the maximum of the data curve it is apparent that curves having their maxima close to this time will give a better fit of their normalized curves. The size and shape of the region over which the fitting takes place, in all probability, can be largely influenced by the normalization procedure.

The contour for the actual data curve is shown in Figure 14. Again the minimum occurs in a valley and a range of rate constants give acceptable fits. As seen earlier, the value of $k_2$ is determined more accurately than the value of $k_1$.

Once $k_1$ and $k_2$ have been determined it is then possible, if the initial concentrations are known from knowledge of the resin system, to calculate curves for the concentration of each species during cure. In this way the absorbance of isocyanate at its maximum can be calculated.

Comparison of results using concentration and normalized absorbance for the experimental data demonstrates the effectiveness of this approach. The concentration procedure results in $k_1$ = 1.85/min and $k_2$ = 0.35 ℓ/mol/min. The normalized absorbance procedure results in $k_1$ = 1.94/min and $k_2$ of 0.33 ℓ/mol/min. This demonstrates the ability of the normalization procedure to accurately determine the rate constants from the shape of the curve. The maximum isocyanate absorbance calculated by the normalization procedure, .712 M, agrees very well with the experimentally determined value of .692 M.

Conclusions
_____

The rate constants for deblocking and cure can be obtained by monitoring the concentration of free isocyanate during cure. In the absence of actual concentration data it is possible to obtain good values for the rate constants using a normalization procedure. In addition to the normalization procedure used in this work it is possible to define other normalization functions which may produce significant changes in the contour surface over which the search occurs.

The use of contour maps provides a good estimate of the uncertainty in optimized rate constants.

In general terms, the Nelder-Mead simplex routine can be used to rapidly generate rate constants from systems which do not give analytical solutions. This could be applied to numerous other systems.

One feature of this approach is the ease with which the differential equations used by the Euler-Romberg integration method can be changed to correspond to different chemical systems. As an example, if the blocked isocyanate contains two different functionalities, it is possible to add equations describing the deblocking and crosslinking reaction of each functionality. Similarly, for resins which have no hydroxyl functionality the same basic program can be used to determine the rate constant for the first order deblocking.

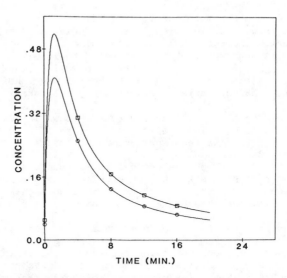

FIGURE 13  UNNORMALIZED TRIAL CURVE FROM CONTOUR PLOT VALLEY
(F <0.005, □ $k_1$ = 1.25/min, $k_2$ = 0.7 ℓ/mol/min TRIAL CURVE;
O $k_1$ = 1.0/min, $k_2$ = 1.0 ℓ/mol/min DATA CURVE)

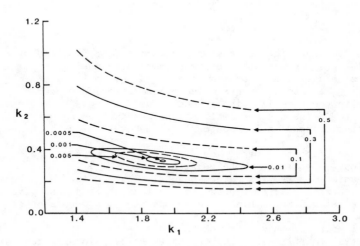

FIGURE 14  CONTOUR PLOT FOR EXPERIMENTAL DATA OF FIGURE 4 USING NORMALIZATION

## Acknowledgments

The authors would like to thank the coworkers who contributed to the thought and programming which went into this work. In particular, C. Mike Neag, Sharon Kaffen, Kirk J. Abbey, Cheng-Yih Kuo, Ann Kah, Ted Niemann, Mark Koehler and Alan Toman are acknowledged.

## Literature Cited

1. Wicks, Z., Prog. Org. Coatings, 1975, 2, 73.

2. Wicks, Z., Prog. Org. Coatings, 1981, 9, 3.

3. Anagnastou, Y., and Jaul, E., J. Coatings Tech., 1981, 53, No. 673, 35.

4. Kordomenos, P. I., Dervan, A. H., and Kresta, J., J. Coatings Tech., 1982, 54, No. 687, 43.

5. Carlson, G. M., Neag, C. M., Kuo, C., and Provder, T., in "Advances in Urethane Science and Technology," Ed. K. C. Frisch, D. Klempner, Technomic Publishing Co., Lancaster Pa., 1984; Vol. 9, p. 47.

6. Carlson, G. M., Neag, C. M., Kuo, C., and Provder, T., In "FT-IR Characterization of Polymers", H. Ishida, Ed., Plenum, New York, 1985; (in press).

7. Provder, T., Neag, C. M., Carlson, G. M., Kuo, C., and Holsworth, R., "Analytical Calorimetry" Ed. P. S. Gill, J. Johnson, Plenum Press, N.Y., 1984; Vol. 5, p. 377.

8. Carlson, G. M., Neag, C. M., Kuo, C., and Provder, T., ACS Polymer Preprints, 1984, 25 (2), 171.

9. L. T. Phai, F. Viollaz, Y. Chamberlin, T. M. Lau, and J. P. Pascault, Makromol. Chem., 1984, 185, 281.

10. Mirgel, V., Proc. XV Congress AFTPV, Cannes, 1982, p. 173.

11. Chien, J. Y., J. Am. Chem. Soc., 1948, 70, 2256.

12. McCalla, T. R., "Introduction to Numerical Methods and FORTRAN Programming", Wiley & Sons, New York, 1967, 341.

13. Olsson, D. M., J. Quality Technology, 1974, 6, 53.

14. Koehler, M. E., Kah, A. F., Neag, C. M., Niemann, T. F., Malihi, F. B., and Provder, T. "Analytical Calorimetry", Ed. P. S. Gill, J. Johnson, Plenum Press, N.Y., 1984; Vol. 5, p. 361.

RECEIVED December 11, 1985

# 22

# Development and Application of Network Structure Models to Optimization of Bake Conditions for Thermoset Coatings

**David R. Bauer and Ray A. Dickie**

**Research Staff, Ford Motor Company, Dearborn, MI 48121**

A physicochemical model of crosslinked network formation has been applied to the evaluation and optimization of cure conditions for thermoset acrylic copolymer/melamine formaldehyde coatings in automotive assembly plant ovens. Cure response has been characterized in terms of an elastically effective crosslink density ($C_{el}$) calculated from a network structure model, a kinetic model of cure, and measured car body temperature profiles. Acceptability of cure has been assessed by comparing the calculated value of $C_{el}$ with the range of values of $C_{el}$ known to result in acceptable paint physical properties. Relationships between oven and paint parameters have been quantified using the model together with an optimization program. It has been found that the minimum bake time to achieve acceptable uniformity of paint cure is principally a function of the cure response (or ~cure window~) of the paint, the maximum allowed air temperature in the oven, and the minimum value of a heat transfer parameter on the car body. At constant bake time, more uniform paint cure can be achieved either by broadening the cure response of the paint or by increasing the minimum heat transfer parameter.

The physical properties of automotive enamels are in large part determined by the crosslink structure developed in the paint films during the baking process. Enamels which are not cured sufficiently (undercured) are generally sensitive to humidity and solvents. In addition, they may be prone to chipping and cold cracking. Paints which have been baked excessively (overcured) exhibit intercoat adhesion failure. That is, subsequent coats

0097-6156/86/0313-0256$06.00/0
© 1986 American Chemical Society

(tutone or repair coats for example) do not adhere well to the coat that is overcured. Clearly, the physical properties of the paint depend on the bake time and temperature. The range of bake times and temperatures over which the paint has acceptable properties defines a ~cure window~. In practice a paint's cure window is determined by measuring a variety of physical properties for a large number of different bake times and temperatures. Such measurements provide useful information but do have several limitations. Although a wide variety of bake conditions are tested, these procedures cannot determine the cure response of a given paint to all the possible time-temperature profiles exhibited by different car bodies and ovens. The results of the physical tests are generally qualitative (e.g., pass or fail) rather than quantitative, and it is not possible to develop a scheme to predict cure response quantitatively for arbitrary bake schedules. Relationships between cure, paint formulation variables, and oven variables are often obscure, and process specifications must be developed for each specific paint.

A network structure model that allows calculation of network parameters from measured extents of reaction has been developed (1) based on the work of Macosko and Miller (2,3). In studies in which calculated network parameters have been compared with physical property measurements on acrylic/melamine coatings, it has been found that for one such parameter, the elastically effective crosslink density ($C_{el}$), there is a well defined range of $C_{el}$ over which acceptable cure is achieved (4). Coatings with values of $C_{el}$ outside this range are either over or undercured. Extents of reaction can be accurately measured for these coatings using infrared spectroscopy (5-9). Based on these studies a kinetic model for crosslinking with hexamethoxymethylmelamine has been proposed (7). In this paper, the kinetic model and the network structure model are combined to provide a method for calculation of cure response for arbitrary bake oven profiles. The cure response of low and high solids paints to typical assembly bake ovens are determined. Oven optimization schemes that use these models together with the nonlinear optimization program SIMPLEX (10) to determine tradeoffs between paint and process parameters are described. In the sections that follow the kinetic and network structure models are described, results on model validation are presented, the important formulation and process variables are discussed, the optimization strategy is described, and finally the results and implications of the models are discussed.

## Model Development and Validation

Kinetic Model. All of the coatings used in this study are hydroxy functional acrylic copolymers crosslinked with melamine formaldehyde crosslinkers. The chemistry of crosslinking with melamine formaldehyde crosslinkers has been discussed in detail elsewhere (5,11). The type and rate of the reactions depend primarily on the structure of the melamine. Fully alkylated melamines (e.g., hexamethoxymethylmelamine) undergo primarily a single reaction: the reaction of alkoxy groups with hydroxy groups on the polymer to form polymer-melamine crosslinks. The reactions of partially alkylated melamines are more complex. Formation of both polymer-melamine crosslinks and melamine-melamine crosslinks occurs. Although extents of reaction of coatings crosslinked with partially alkyated melamines have been measured (5,6), no kinetic model that adequately describes the kinetics of crosslinking of partially alkylated melamines has been developed. A kinetic model of crosslinking in hexamethoxymethylmelamine based on the mechanism proposed by Blank (11) has been developed by Bauer and Budde (7). The mechanism is as follows:

$$NCH_2OCH_3 + H^+ \underset{k_{-1}}{\overset{k_1}{\rightleftharpoons}} \overset{H^+}{NCH_2OCH_3} \quad \text{fast}$$

$$\overset{H^+}{NCH_2OCH_3} \underset{k_{-2}}{\overset{k_2}{\rightleftharpoons}} NCH_2^+ + CH_3OH \quad \text{slow}$$

$$NCH_2^+ + ROH \underset{k_{-3}}{\overset{k_3}{\rightleftharpoons}} \overset{H^+}{NCH_2OR} \quad \text{fast}$$

$$\overset{H^+}{NCH_2OR} \underset{k_{-4}}{\overset{k_4}{\rightleftharpoons}} NCH_2OR + H^+ \quad \text{fast}$$

Analysis assuming steady-state kinetics of the above reactions results in the following expression for the rate of disappearance of polymer hydroxy functionality:

$$[ROH] = \frac{-[H^+]K^{\sim}\{k_3k_2[ROH][NCH_2OCH_3] - k_{-2}k_{-3}[CH_3OH]([ROH^o]-[ROH])\}}{k_3[ROH] + k_{-2}[CH_3OH]} \quad (1)$$

Several assumptions were made in order to analyze kinetic data in terms of this expression (7). First it was assumed that $k_{-2} = k_3$, $k_2 = k_{-3}$, and $k_1/k_{-1} = k_{-4}/k_4$ ( $= K^{\sim}$). Second it was assumed that the rate constants were independent of the extent of reaction i.e., that all six functional groups were equally reactive and that the reaction was not diffusion controlled. The concentration of polymer hydroxyl functionality was determined experimentally using infrared spectroscopy as described elsewhere (7). A major unknown is the instantaneous concentration of methanol. Fits to the kinetic data were made with a variety of assumptions concerning the methanol concentration. The best fit was achieved by assuming that the concentration of methanol was initally constant but decreased at a rate proportional to the concentration of residual polymer hydroxy groups towards the end of the reaction. As discussed elsewhere (7), this is physically reasonable since at high extents of reaction, the rate of consumption of polymer hydroxyl decreases; thus the rate of production of methanol also decreases and the concentration of methanol in the film can be expected to decrease as well. The values of the parameters describing methanol concentration depended on the ratio of methoxy to hydroxy. In the coatings examples that follow, the ratio of methoxy to hydroxy was 1.5 to 1; for this ratio, the best fit to the kinetic data was obtained by assuming that the methanol concentration remained constant up to a hydroxy conversion of 67%. (Typically, values of the methoxy to hydroxy ratio range from 1.5 to 2.5. The value of 1.5 was used in this work in order to minimize the effects of side reactions such as condensation of leftover methoxy groups. The effect of this restriction on the calculated results is, however, minor.) With these assumptions, Equation 1 can be integrated to yield, for $p_A = 0$ to 0.67:

$$Kt = -0.5 \ln(p_A^2 - 2.64p_A + 1.49) + 0.17 \ln[(1.635 - 2p_A)/(3.645 - 2p_A)] + 0.34 \quad (= SK) \qquad (2a)$$

and for $p_A = 0.67$ to 1.00:

$$Kt = -\ln(1.49 - 1.45p_A) + 0.22 \quad (= SK) \qquad (2b)$$

where $p_A$ is the fractional extent of reaction of hydroxy groups, $K = [H^+] \tilde{K} k_2$, and t is the reaction time. The fit of Equation 2 to the experimentally determined extent of reaction of polymer hydroxyl groups as a function of time (at constant temperature) is shown in Figure 1a. From this data, the value of the rate constant K can be determined. K has been measured as a function of temperature and found to fit an Arrhenius expression with an activation energy of 12.5 Kcal/mole (7). A comparison of experimental and calculated extent of reaction versus temperature (at constant reaction time) is shown in Figure 1b. The rate constant K is proportional to catalyst level (7) and thus can be adjusted by adjusting the catalyst level so that optimum cure is achieved under standard fixed bake conditions. Considering the assumptions in the model, the agreement between experimental and predicted extents of reaction is excellent. The fits in Figure 1 are to the idealized conditions of either constant bake temperature or constant bake time. The extent of reaction for arbitrary bake profiles can be determined from Equation 2 and the temperature dependence of K. In the present study, the extent of reaction has been determined using an incremental isothermal state approximation: the bake profile has been broken into a sequence of constant temperature intervals, K has been calculated for each interval, the values of K t have been summed over the bake cycle to yield a dimensionless parameter SK, and this parameter and Equation 2 have been used to determine $p_A$. The calculation of K t

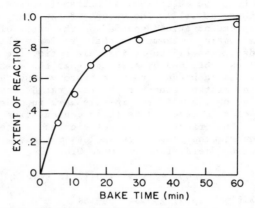

Figure 1a. Experimental (O) and calculated (———) extent of reaction versus bake time at 125 C and a PTSA concentration of $3.8 \times 10^{-3}$ M.

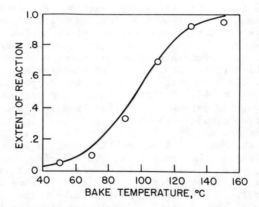

Figure 1b. Experimental (O) and calculated (———) extent of reaction versus bake temperature at a constant bake time of 20 min and a PTSA concentration of $5.8 \times 10^{-3}$ M.

can of course also be handled as an integration over time if an appropriate representation of the temperature history as a function of time is available.

Network Structure Model. The extent of reaction together with the hydroxy equivalent weight determines the number of chemical crosslinks formed during cure. The extent of reaction does not, however, correlate well with such physical measurements of cure as solvent resistance and intercoat adhesion (1). It is found experimentally that coatings based on low molecular weight resins ("high solids" coatings) require higher levels of chemical crosslinking than do conventional coatings based on higher molecular weight resins ("low solids" coatings) (1). The main reason for this is that a large fraction of the chemical crosslinks formed in the high solids coating are simply builiding up the molecular weight of the starting materials. In order to develop a parameter which correlates with physical measures of cure it is necessary to devise a counting scheme which counts only "effective" crosslinks. Such schemes have long been of interest in, for example, studies of rubber elasticity; a network chain is considered to be elastically effective if its ends are effective crosslinks (i.e., crosslinks from which three or more chains lead out to the infinite network, to the sample surface) (3). The elastically effective crosslink density $C_{el}$ has been found to correlate well with physical measures of paint cure (1,4). Calculation of $C_{el}$ is based on the probabilistic network model of Miller and Macosko (2,3); expressions have been derived for calculation of $C_{el}$ in coatings crosslinked with fully and with partially alkylated melamines. Only the model for the fully alkylated melamine is used here. Since the model has been described in detail elsewhere (1), only a summary is presented here.

The model requires as input the extent of reaction and the distribution of functionality on the polymer and crosslinker. The extent of reaction used in the network structure model can be measured or calculated from a kinetic model. It is assumed that the melamine crosslinker is six functional and that all six groups are equally reactive; ring formation is ignored. The distribution of hydroxy functionality on the polymer is calculated for a polymer of given molecular weight and hydroxy equivalent weight assuming random copolymerization and a most probable molecular weight distribution. (The calculation could certainly be generalized to account for both non random polymerization and different molecular weight distributions, but this was not necessary in the present study as the molecular weight distributions of the polymers of interest were very close to most probable.) The results are expressed in terms of the number fraction of polymers with I hydroxy groups on the chain, $X_n(I)$. In practice, the value of $X_n(I)$ is most important for low values of I since polymers with low functionality have the highest proportion of ineffective crosslinks. The Macosko formalism calculates the probability that looking out from a given group, Z, is a finite chain. This probability is denoted $P(F_Z^{out})$. If A denotes hydroxy groups on the polymer, B denotes methoxy groups on the melamine, and ALK is the ratio of methoxy to hydroxy, then the following expressions can be derived for $P(F_A^{out})$ and $P(F_B^{out})$:

$$P(F_A^{out}) = 1 - p_A + p_A P(F_B^{out})^5 \qquad (3a)$$

$$P(F_B^{out}) = 1 - p_A/ALK + (p_A/ALK)(1/X)\sum_{I=1} IX_n(I)P(F_A^{out})^{I-1} \qquad (3b)$$

where $X = IX(I)$. A computer program using Newton's method has been written to solve for $P(F_A^{out})$ and $P(F_B^{out})$. For a crosslink to be elastically effective, there must be at least three independent paths out to the infinite network. The quantity $1 - P(F_Z^{out})$ is the probability that a single group on Z leads out to the infinite network. To calculate $C_{el}$ it is necessary to modify this probability by the probabilty that there are at least two other paths out to the infinite network. This is just 1 - (the probability that there is either no other path or that there is only one additional path). Thus, $C_{el}$ is given by the following expression:

$$C_{el} = (1/2Q)\{[1 - P(F_A^{out})](1/X)\sum_{I=3} IX_n(I)\{1 - P(F_A^{out})^{I-1} - (I-1)[1-P(F_A^{out})]P(F_A^{out})^{I-2}\}$$
$$+ ALK[1-P(F_B^{out})][1-P(F_B^{out}) - 5[1-P(F_B^{out})]P(F_B^{out})^4]\} \qquad (4)$$

where Q is the hydroxy equivalent weight of the resin (including crosslinker); this is a special case of Equation A6 of Ref. 1, with g = 6 and h = 0.

$C_{el}$ has been found to correlate with physical measures of cure for a variety of coatings independent of polymer molecular weight and melamine crosslinker type. There exists a range of $C_{el}$ around an optimum value of $1.0 \times 10^{-3}$ moles/g for which acceptable cure is achieved (4). If $C_{el}$ is below about $0.8 \times 10^{-3}$ moles/g, the coatings tend to be humidity sensitive or otherwise undercured. As $C_{el}$ decreases from $0.8 \times 10^{-3}$, underbake problems become increasingly severe. Coatings with $C_{el}$ above $0.8 \times 10^{-3}$ do not exhibit underbake problems. Coatings whose value of $C_{el}$ is greater than $1.25 \times 10^{-3}$ moles/g typically exhibit intercoat adhesion failure. The boundary between coatings acceptably cured and overbaked coatings is remarkably sharp (4). All coatings studied with a value of $C_{el}$ greater than $1.27 \times 10^{-3}$ exhibited severe intercoat adhesion failure while those with a value less than $1.24 \times 10^{-3}$ had no failure.

Values of $C_{el}$ can be calculated for arbitrary bake histories for arbitrary coatings by combining the kinetic model with the network structure model. Four coatings were considered in the current study: one, a conventional low solids paint formulated with a butylated melamine crosslinker, serves essentially as a control for evaluation of three hypothetical high solids formulations based on hexamethoxymethylmelamine. The kinetic model was developed for crosslinking with hexamethoxymethylmelamine and is not directly applicable to the butylated melamine crosslinker. There is no practical kinetic model for crosslinking with this particular melamine currently available. In order to calculate $C_{el}$ for this coating as a function of bake history, the parameters used in the kinetic and network structure models were modified empirically. Extents of reaction for the low solids coating were measured spectroscopically as a function of bake temperature at constant bake time. These extents of reaction were used in the general network structure model of Ref. 1 to determine experimental values of $C_{el}$ for this coating. The results are shown in Figure 2. The value of the activation energy was modified in the kinetic model. In the network structure model, the value assumed for the functionality of the melamine was increased (to reflect the fact that butylated melamines are polymerized) and the assumed hydroxy equivalent weight was adjusted so that the calculated values of $C_{el}$ matched as closely as possible the experimental values. The values of the parameters used are given in Table I.

Table I. Paint Parameters

| Parameter | Low Solids | High Solids 1 | High Solids 2 | High Solids 3 |
|---|---|---|---|---|
| Acrylic $M_n$ | 8500 | 4000 | 3000 | 2000 |
| Melamine Functionality | 10 | 6 | 6 | 6 |
| Methoxy:hydroxy | 1.5 | 1.5 | 1.5 | 1.5 |
| Q | 760 | 750 | 700 | 650 |
| Activation Energy | 8 | 12.5 | 12.5 | 12.5 |
| Cure Window | 50 | 40 | 35 | 30 |

Table II. Model Verification for Low Solids Coating

| Body Position | $C_{el} \times 10^3$ moles/g | |
|---|---|---|
| | Experimental | Model |
| Center Roof | 1.09 | 1.11 |
| Center Hood | 1.07 | 1.05 |
| Fender | 1.10 | 1.06 |
| B Post | 0.97 | 1.01 |
| Rocker Panel | 0.98 | 0.97 |

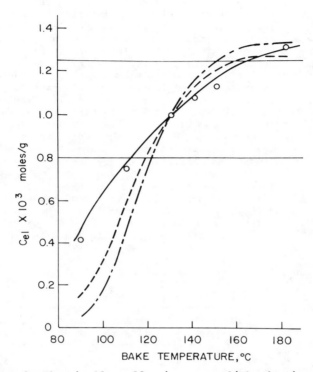

Figure 2. Elastically effective crosslink density versus bake temperature for a 17 minute bake. Low solids: O experimental values, ——— model values; High solids 1: -----; High solids 3: -·-·-·

The agreement between the experimental and calculated values of $C_{el}$ is excellent. The data shown in Figure 2 are for a constant bake time of 17 minutes. The upper and lower limits on $C_{el}$ define a cure window. The cure window for the low solids coating is 50 C. The model was further tested by measuring extents of reaction and temperature profiles for samples attached to different parts of a car body which passed through a pilot plant oven. This simulation tested the model under conditions where the substrate temperatures were far from constant. As shown in Table II, the agreement between the experimental and calculated values of $C_{el}$ is again excellent.

As has been discussed elsewhere, high solids coatings tend to have narrower cure windows than low solids coatings (4). They also tend to be more sensitive to batch-to-batch variations which tends to further narrow the cure window. For this study, coating parameters were chosen to give cure windows at constant bake time of from 30 to 40 C (see Figure 2). All coatings were assumed to be crosslinked with hexamethoxymethylmelamine with a methoxy to hydroxy ratio of 1.5. The rate constant K (catalyst level) was adjusted so that an extent of reaction of 85% was acheived after 17 minutes at 130 C. The cure window was adjusted by varying the molecular weight of the polymer (lowering the molecular weight reduces the cure window). The hydroxy equivalent weight was adjusted so that $C_{el} = 1.00 \times 10^{-3}$ at 85% conversion. The parameters used are summarized in Table I. This choice of coatings -- though admittedly somewhat arbitrary -- allows the effect of narrowing the cure window on paint cure uniformity, and hence on process parameters, to be determined independently of details of paint chemistry and formulation.

## Cure in Assembly Ovens: Heat Transfer

Typical studies of cure kinetics in the laboratory involve measurements of cure at constant temperature and constant time (for example Figure 2). Cure of painted car bodies in assembly ovens is considerably more complex. The cross section of a typical assembly oven is shown in Figure 3. Oven air temperatures can be controlled independently in several zones (3 - 5). The complex airflow patterns lead to variations of air temperature over the cross section as well as down the length of the oven. Paint cure is controlled by the substrate surface temperature rather than the air temperature. Thus heat transfer between the air and the surface of the car body is important. Variables affecting the rate of heat transfer include air velocity and substrate size, shape, and composition. Typically, exposed sheet metal heats up rapidly while more massive metal components such as pillars heat up more slowly. Thus heat transfer parameters vary from place to place on a given car body, from car body to car body, and from oven to oven. Assuming that local temperature and air velocity changes are small during a short time interval t, a general expression for the transient heating of a conductive substrate in a hot air stream may be written in the following form:

$$T(t+\Delta t) = Ta - [Ta - T(t)]e^{-Z \Delta t} \qquad (5)$$

where $T(t)$ is the car body temperature at time t, Ta is the oven air temperature, and Z is a heating rate constant. Data relating car body temperatures to oven air temperatures are available from routine checks on oven performance in which car bodies with thermocouples attached are passed through the ovens. Typical thermocouple locations are shown in Figure 4. From this data and Equation 5, the heating rate constant can be derived for different car body positions. Due mainly to the fact that oven air temperatures vary somewhat over the crossection of the oven, the standard deviation of each individual measurement of Z is rather large. Nevertheless, the data are reasonably consistent from run to run. It is found that each car body has a range in heating rate constants. The minimum heating rate constant, $Z_{min}$, varies from 0.13 - 0.2 $min^{-1}$ while the maximum heating rate constant, $Z_{max}$, varies from 0.45 - 0.65 $min^{-1}$.

The heating rate constants determine the response of the car body to oven air temperature. Even if the oven air temperature was constant, different parts of the car would see different temperature profiles since they have different heating rate constants. The car body part with the smallest heating rate constant would be the part that would tend to be underbaked while the part with the highest heating rate constant would tend to be overbaked. There are several other process variations which can affect cure response. For example, occasionally it is necessary to stop the conveyor with cars still in the oven. If the stoppage is at all lengthy, the oven temperature is reduced. If the stoppage is short, no attempt can be made to reduce temperatures. Calculations of the effect of such stoppages on final crosslink density for typical temperature profiles reveals that the effects are small for the control coating. A more important variation is that of multiple bakes. An enamel coating will receive more than one bake if it is part of a tutone car or if it needs repair. Thus different coats on the same car can receive different numbers of bakes. This places a severe constraint on the cure window of the paint since the paint must achieve a crosslink density of at least $0.8 \times 10^{-3}$ moles/g after a single bake at car body locations with a low heating rate constant and yet must not be overbaked (i.e., $C_{el}$ less than $1.25 \times 10^{-3}$ moles/g) after several bakes. It is known that the control coating passes all required performance specifications even after three passes (a worst case) through typical ovens. Calculations using the modified kinetic model yield values for $C_{el}$ that are generally less than $1.25 \times 10^{-3}$ moles/g after two passes but that are slightly greater than this value after three passes. Considering the many assumptions of the calculation, this does not seem a serious discrepancy. For the

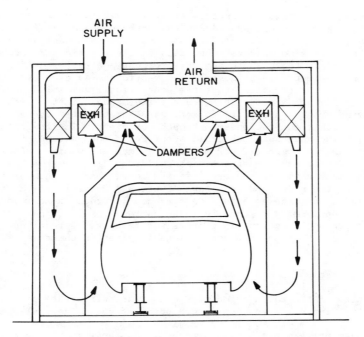

Figure 3. Cross section of typical assembly paint bake oven.

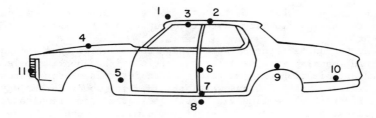

1. HIGH AIR
2. CENTER ROOF
3. SIDE ROOF
4. CENTER HOOD
5. FENDER
6. B POST
7. ROCKER PANEL
8. LOW AIR
9. UNDER WHEEL HOUSE AT VINYL SEALER (METAL)
10. QUARTER PANEL (2-IN ABOVE VINYL SEALER)
11. CENTER OF GRILL (PLASTIC)

Figure 4. Location of thermocouples for paint bake oven survey.

purposes of the model, the value of $C_{el}$ calculated after two passes through the oven was used to evaluate overbake.

The car body temperature data used to calculate heating rate constants were also used as input data in the kinetic and network structure models to evaluate the cure response of the different paints to current assembly ovens. It was found (as expected from the physical property data) that the conventional low solids coating generally had acceptable cure performance. That is, there were few underbakes predicted for one pass through the ovens and few over bakes for two passes through the ovens. The few calculated over and underbakes that did occur were not severe (i.e., did not deviate greatly from the maximum and minimum allowed crosslink densities). When the high solids coatings were tested with the same data set, however, it was found that the number and severity of the under and over bakes increased dramatically as the cure window of the coating decreased. In order to achieve an acceptable uniformity of cure in these narrow cure window coatings, some modification in process was found to be required. A discussion of the tradeoffs and options is given in the next section.

Oven Optimization

The results of the calculation of $C_{el}$ for the different coatings using the typical oven temperature profiles raises several questions:
o Can oven performance be improved by adjustments in the oven air temperature?
o Are there specific oven strategies which would improve performance?
o Quantitatively what are the tradeoffs between cure window, paint cure uniformity, oven throughput, and heating rate constant?

In order to answer these questions, the kinetic and network structure models were used in conjunction with a nonlinear least squares optimization program (SIMPLEX) to determine cure response in "optimized ovens". Ovens were optimized in two different ways. In the first the bake time was fixed and oven air temperatures were adjusted so that the crosslink densities were as close as possible to the optimum value. In the second, oven air temperatures were varied to minimize the bake time subject to the constraint that all parts of the car be acceptably cured. Air temperatures were optimized for each of the different paints as a function of different sets of minimum and maximum heating rate constants.

In general the SIMPLEX program adjusts variables subject to supplied constraints to minimize a given object function. The object function must be an explicit function of the variables. In the case where paint cure uniformity was optimized, the oven was divided into three zones and the variables were the three oven zone air temperature settings. Temperatures within the zone were interpolated from the two nearest zone settings. It was assumed that the oven air temperature was constant across a cross section of the oven. The oven air temperatures were constrained to be less than a given maximum air temperature. The maximum is in practice chosen to prevent degradation of paint or plastic components passing through the oven. Unless otherwise noted, the maximum air temperature was 149 C. For a given paint there is a direct relationship between crosslink density, extent of reaction, and the sum of K t (SK). Thus minimizing the difference in crosslink density is the same as minimizing the difference in SK. For each paint there are values of SK that correspond to the minimum and maximum acceptable values of $C_{el}$ ($SK_{min}$ and $SK_{max}$) and to the optimum value of $C_{el}$ ($SK_{opt}$). These values are of course different for each paint. For a given set of air temperatures, the lowest value of SK ($SK_{low}$) is determined using $Z_{min}$ to calculate body temperatures while the highest value of SK ($SK_{high}$) is determined using $Z_{max}$. The value of $SK_{high}$ is calculated assuming two passes through identical ovens. Paint cure quality is maximized when $SK_{low}$ and $SK_{high}$ are as close as possible to $SK_{opt}$. The specific object function that was used was:

$$OBJ = (SK_{low} - SK_{opt})^2 + (SK_{high} - SK_{opt})^2 \qquad (6)$$

In order to insure that there were as few over and under bakes as possible, the object function was modified so that it increased rapidly when $SK_{low}$ was less than $SK_{min}$ or when $SK_{high}$ was greater than $SK_{max}$.

The results of this optimization are shown in Figure 5. Differences in $C_{el}$ of less than $0.45 \times 10^{-3}$ moles/g are acceptable while differences greater than that imply that a part of the car body is either undercured, overcured, or both. The control low solids coating is found to achieve acceptable cure in optimized ovens as long as $Z_{min}$ is greater than 0.12 min$^{-1}$. Comparison of the paint cure uniformity of the low solids control in optimized ovens with actual assembly plant data on ovens with similiar values of $Z_{min}$ reveal that although the optimized oven does have better cure uniformity, the differences are not great. For example, for a $Z_{min}$ of 0.2 min$^{-1}$, the assembly oven achieved a $\Delta C_{el}$ of 0.35 versus $0.3 \times 10^{-3}$ moles/g for the optimized oven. In addition, analysis of the optimized air temperatures revealed no specific strategy to optimize oven performance aside from carefully controlling average air temperatures. Values of $\Delta C_{el}$ decrease with increasing cure window and increasing $Z_{min}$. Acceptable paint cure uniformity was achieved in a 17 minute bake for the paint with a cure window of 35 C only if $Z_{min}$ was greater than 0.2 min$^{-1}$. The results are not very sensitive to the value of $Z_{max}$. Values of $\Delta C_{el}$ increase with decreasing cure window because a given variation in temperature profiles results in a wider disparity in crosslink density for the narrower cure window paints.

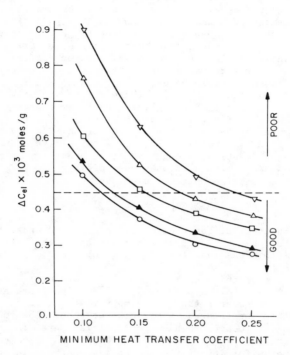

Figure 5. Paint cure uniformity versus minimum heating rate constant for the low solids paint (◯), high solids 1 (□), high solids 2 (△), and high solids 3 (▽). The open symbols are for a 17 minute bake and the closed symbols are for a 25 minute bake.

Values of $\Delta C_{el}$ decrease with increasing $Z_{min}$ since the car body temperature profiles become more uniform. Another way to increase car body temperature uniformity (and thus decrease $\Delta C_{el}$) is to increase the bake time. For example, increasing the bake time from 17 minutes to 25 minutes essentially compensates for a decrease in cure window of 50 C to 35 C.

The tradeoff between bake time and cure performance can be seen more directly by holding paint cure uniformity constant while minimizing the bake time. Again the variables are the oven zone air temperatures. In this case bake time is another variable. The constraints are that $SK_{low}$ be greater than or equal to $SK_{min}$ and that $SK_{high}$ be less than or equal to $SK_{max}$. Since these are not specific constraints on the air temperatures, they must be incorporated into the object function. This is fairly straightforward once it is recognized that minimum bake times are achieved when $SK_{low}$ is as close as possible to $SK_{min}$ and $SK_{high}$ is as close as possible to $SK_{max}$. Put another way, minimum bake times are achieved when $\Delta C_{el}$ is less than, but as nearly equal as possible to, $0.45 \times 10^{-3}$ moles/g.

The results of the minimization of bake time are shown in Figure 6. There is a direct tradeoff between bake time, cure window and $Z_{min}$. Narrowing the cure window increases the minimum bake time required for acceptable cure uniformity. Increasing $Z_{min}$ allows the bake time to be shorter. Narrow cure window paints require more uniform car body temperatures; this can only be achieved by increasing the bake time or $Z_{min}$. Somewhat surprisingly, the constraint on maximum air temperature affects the minimum bake time only for paints with wide cure windows and for car bodies with relatively high values of $Z_{min}$. Considering the typical range of $Z_{min}$ normally encountered, a bake time of 17 minutes is just long enough to provide acceptable paint cure for the conventional low solids enamel. It is possible to improve oven productivity (i.e., shorten the bake time) for this coating by increasing the lowest value of $Z_{min}$ and raising the maximum allowed air temperature. For example, if oven air flows could be adjusted so that $Z_{min}$ was at least $0.2$ min$^{-1}$, the bake time could be reduced to 14 minutes while still maintaining cure quality. A further reduction to 12 minutes would be possible if the maximum air temperature could be increased to 166 C. Coatings with narrower cure windows than the control can not be acceptably cured even if the air temperatures are optimized. Longer bake times or improvements in heat transfer would be required to acceptably cure the narrow cure window paints. The tradeoff between paint cure window and oven productivity is shown in Figure 7. Assuming a fixed oven length so that longer bake times require slower line speeds, it can be seen that a small decrease in paint cure window can be tolerated without loss of productivity if the minimum heating rate is high. If, however, the minimum heating rate is low, any narrowing of the cure window results in a loss of oven productivity.

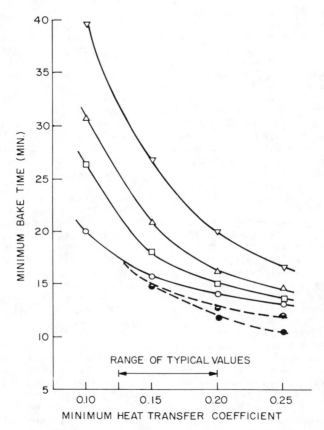

Figure 6. Minimum bake time versus minimum heating rate constant for the low solids paint (○), high solids 1 (□), high solids 2 (△), and high solids 3 (▽). The open symbols are for a maximum air temperature of 149 C. The half filled circles are for a maximum air temperature of 157 C and the filled circles are for a maximum air temperature of 166 C.

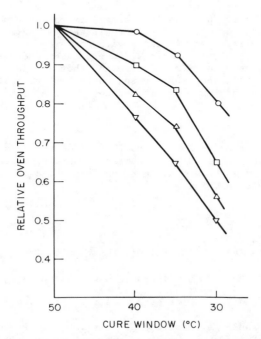

Figure 7. Relative oven throughput versus paint cure window for values of $Z_{min}$ of 0.25 ($\bigcirc$), 0.2 ($\square$), 0.15 ($\triangle$), and 0.10 $min^{-1}$ ($\triangledown$).

Conclusion

A network structure model has been developed from which a parameter that correlates well with physical measures of paint cure can be calculated. This model together with a kinetic model of crosslinking as a function of time and temperature has been used to evaluate the cure response of enamels in automotive assembly bake ovens. It is found that cure quality (as measured by the number and severity of under and overbakes) is good for a conventional low solids enamel. These results are in agreement with physical test results. Use of paints with narrower cure windows is predicted to result in numerous, severe under and over bakes. Optimization studies using SIMPLEX revealed that narrow cure window paints can be acceptably cured only if the bake time is increased or if the minimum heating rate on the car body is increased.

Acknowledgments

This paper is dedicated to Dr. Pierre Thirion on the occasion of his retirement. The authors wish to acknowledge the assistance of Fred Oblinger in measuring the heating rate constants and Joe Norbeck in applying the SIMPLEX program.

Literature Cited

1. D. R. Bauer and G. F. Budde, Ind. Eng. Chem., Prod. Res. Dev., 20, 674 (1981).
2. C. W. Macosko and D. R. Miller, Macromolecules, 9, 199 (1976).
3. D. R. Miller and C. W. Macosko, Macromolecules, 9, 206 (1976).
4. D. R. Bauer and R. A. Dickie, J. Coat. Technol., 54 (685), 57 (1982).
5. D. R. Bauer and R. A. Dickie, J. Polym. Sci., Polym. Phys., 18, 1997 (1980).
6. D. R. Bauer and R. A. Dickie, J. Polym. Sci., Polym. Phys., 18, 2015 (1980).
7. D. R. Bauer and G. F. Budde, J. Appl. Poly. Sci., 28, 253 (1983).
8. J. Dorffel and U. Biethan, Farbe und Lacke, 82, 1017 (1976).
9. M. G. Lazzara, J. Coat. Technol., 56 (710), 19 (1984).
10. J. P. Chandler, Quantum Chemical Program Exchange Program No. 307 (SIMPLEX), 1965.
11. W. J. Blank, J. Coat. Technol., 51 (656), 61 (1979).

RECEIVED November 14, 1985

# 23

# A Kinetic Study of an Anhydride-Cured Epoxy Polymerization

**C. C. Lai, Delmar C. Timm, B. W. Eaton, and M. D. Cloeter**

**Department of Chemical Engineering, University of Nebraska, Lincoln, NE 68588-0126**

>  Dynamic data are simulated utilizing a kinetic model descriptive of a thermoset, step-growth polymerization. Monomeric concentrations and the cumulative molar concentration of polymeric species fit experimental data, yielding estimates of model parameters and constants. To fit population density distribution dynamics of oligomeric species contained within the sol fraction, kinetic rate constants must be dependent on chemical functionality, viscosity plus steric hindrance factors. Numerical integrations utilized a Runge-Kutta-Gill algorithm coupled to a least squares objective function.

Knowledge of the fundamentals which control a polymerization cure are paramount to developing strategies to optimize a materials physical performance. Extensive kinetic studies have been reported on thermoplastic resins, particularly in the regions of initial rates of polymerization and in dilute solutions. Considerably less work has been published at high conversions for bulk polymerizations. The present research initiates a long-term objective to comprehensively model bulk polymerizations of thermosetting resins. The degree of network development within the resin will ultimately be extensive, with the average molecular weight between crosslinks approaching the order of 100. Goals include the demonstration of procedures for determining reaction rates as functions of the environment and the development of strategies which will optimally control crosslink average molecular weight, the distribution of crosslink average molecular weights, and the average molecular weight of the network megamolecules.

The resin system selected to initiate these studies is a step-growth anhydride cured epoxy. The approach to the kinetic analysis is that which is prevalent in the chemical engineering literature on reactor design and analysis. Numerical simulations of oligomeric population density distributions approximate experimental data during the early stages of the cure. Future research will extend the simulation to higher extents of reaction. Prior research (1-5) provides a foundation based on a statistical approach, but model constraints do not incorporate such factors as steric hindrance or viscosity controlled kinetic rates. The current research shows that such factors are necessary even prior to gelation for this step growth polymerization.

## Reagents

Shell's Epon 828, a blend of oligomers of diglycidyl ether of bisphenol A (DGEBA, n=0), was used.

$$CH_2CHCH_2(OC_6H_4C(CH_3)_2C_6H_4OCH_2CHCH_2)_nOC_6H_4C(CH_3)_2C_6H_4OCH_2CHCH_2$$
$$\underset{O}{\phantom{CH_2CH}} \phantom{CH_2(OC_6H_4C(CH_3)_2C_6H_4OCH_2}\underset{OH}{\phantom{CHCH_2}} \phantom{)_nOC_6H_4C(CH_3)_2C_6H_4OCH_2CH}\underset{O}{\phantom{CH_2}}$$

The resin is more than 90% DGEBA with the remaining material predominately at n=1. The curing agent, nadic methyl anhydride (NMA), is manufactured by Allied Chemical. The catalyst was benzyl dimethylamine (BDMA). All materials were stored in a desiccator to minimize moisture pickup.

## Curing Reactions

Tanaka and Kakiuchi (6) proposed catalyst activation via a hydrogen donor such as an alcohol as a refinement to the mechanism discussed by Fischer (7) for anhydride cured epoxies in the presence of a tertiary amine. The basic catalyst eliminates esterification reactions (8). Shechter and Wynstra (9) further observed that at reaction conditions BDMA does not produce a homopolymerization of oxiranes.

The chemistry of the cure is as follows:

$$ROH + NR_3 \underset{\leftarrow}{\overset{K_1}{\rightarrow}} RO^- \ HNR_3^+$$

$$RO^- \ HNR_3^+ + CR_ACO \overset{k_1,j}{\rightarrow} ROCR_ACO^- \ HNR_3^+ \underset{\leftarrow}{\overset{K_2}{\rightarrow}} ROCR_ACOH + NR_3$$
$$\phantom{RO^- \ HNR_3^+ + C}\underset{O}{R_A}\underset{O}{\phantom{CO}} \phantom{\overset{k_1,j}{\rightarrow} ROC}\underset{O}{R_A}\underset{O}{\phantom{CO^-}} \phantom{HNR_3^+ \underset{\leftarrow}{\overset{K_2}{\rightarrow}} ROC}\underset{O}{R_A}\underset{O}{\phantom{COH}}$$

$$ROCR_ACO^- \ HNR_3^+ + H_2C \ CHR_E \overset{k_2,j}{\rightarrow} ROCR_ACOCHCH_2O^- \ HNR_3^+ \underset{\leftarrow}{\overset{K_1}{\rightarrow}}$$

$$ROCR_ACOCHCH_2OH + NR_3$$

Polymeric molecules contain reactive hydrogen sites in the form of alcohols and carboxylic acids. The reaction sequence, alcohol plus anhydride and acid plus oxirane, results in an increase of one in the degree of polymerization if the oxirane is DGEBA. Polymeric species also supply oxiranes via the pendant $R_E$. These reactions are generalized by the notation:

$$A_j + C \underset{\leftarrow}{\overset{K_1}{\rightarrow}} A_jC \tag{1}$$

$$A_jC + A \overset{k_{1,j}}{\rightarrow} A_jAC \underset{\leftarrow}{\overset{K_2}{\rightarrow}} A_jA + C \tag{2}$$

$$A_jAC + E \overset{k_{2,j}}{\rightarrow} A_{j+1}C \underset{\leftarrow}{\overset{K_1}{\rightarrow}} A_{j+1} + C \tag{3}$$

$$A_jAC + A_k \overset{k_{2,j}}{\rightarrow} A_{j+k}C \underset{\leftarrow}{\overset{K_1}{\rightarrow}} A_{j+k} + C \tag{4}$$

$$A_jAC + A_kA \overset{k_{2,j}}{\rightarrow} A_{j+k}CA \underset{\leftarrow}{\overset{K_1}{\rightarrow}} A_{j+k}A + C \tag{5}$$

Equation 1 expresses a state of equilibrium between an alcohol $A_j$ on a molecule whose degree of polymerization is j, the catalyst C and the alkoxide anion $A_jC$. In Relation 2 this activated intermediate reacts with monomeric anhydride A, forming an acid adduct $A_jAC$, which dissociates, forming an unassociable carboxylic acid $A_jA$.

Reactions 3-5 depict the union of a carboxylic intermediate with a monomeric epoxide E, or with pendant oxiranes on macromolecules containing at least one hydroxyl group, $A_k$, or one carboxyl group, $A_kA$. Reaction 4 yields a molecule that contains at least two hydroxyls. Reaction 5 describes a product molecule that contains at least one alkoxide anion and one carboxyl group. The activated intermediate involved in Reaction 5 is therefore described by $A_{j+k}CA$. The $A_jC$ segment is the alkoxide anion; the segment $A_kA$ represents the carboxylic acid. The notation explicitly describes the end group participating in the reaction, but the molecule's functionality is implicit. The intent is to identify a macromolecule with a high functionality but once, either as $A_k$ or as $A_kA$.

The rate constant $k_{i,j}$ will account for the number of polymerization sites per molecule. The initial index on the rate constant $k_{i,j}$ describes the chemical moiety of the catalyst complex. The second index describes the molecular weight of the reactive adduct. The equilibrium constant $K_i$ is similarly defined. If i=1, an alcohol group is singled out, if i=2, a carboxylic acid participates in the reaction. The total number of reactive sites is invariant during a batch cure, but each coupling of macromolecules (see Equations 4 and 5) yields a specie that accumulates these hydrogen sites. If each site has an equal probability of reacting, a molecule that contains multiple sites will, on the average, experience reactions that are proportional to its functionality, which in turn is a function of the molecular weight of the molecule. In the present work, the catalyst concentration is nearly equal to the concentration of reactive sites.

Kinetic Reaction Modeling

If the alcohol and acid complexes are at equilibrium, their respective molar concentrations may be expressed by

$$(A_jC) = K_1 \, A_j \, C \tag{6}$$

$$(A_jAC) = K_2 \, A_jA \, C \tag{7}$$

For batch, isothermal polymerizations, the principle of conservation of population yields for alcohol adducts

$$d(A_jC)/dt = -k_{1,j} \, A \, (A_jC) + k_{2,j-1} \, E \, (A_{j-1}AC) + \sum_{n=1}^{j-1} k_{2,n} \, (A_nAC)P_{j-n} / 2 \tag{8}$$

The several kinetic rates are a consequence of Reactions 2-5. The first two represent monomeric additions; the third describes all rates by which two polymeric molecules smaller than j form a molecule of size j. The molecule $A_nAC$ contains a carboxylic anion; the second molecule $P_{j-n}$ supplies an oxirane and, in general, is any polymeric molecule of size j-n.

$$P_j = A_j + A_jA \tag{9}$$

Equation 10 denotes reactions involving acid adducts.

$$d(A_jAC)/dt = k_{1,j} \, A \, (A_jC) - k_{2,j} \, (A_jAC) \, [E + P_{TOT}] = 0 \tag{10}$$

Monomeric additions result in the first and second rate expressions. Polymeric molecules react with the activated intermediate at a rate that is proportional to their cumulative molar concentration $P_{TOT}$.

$$P_{TOT} = \Sigma \, P_j \tag{11}$$

The rate of anhydride addition, see Equation 10, is also the rate limiting step.

The addition of Equations 8 and 10, subject to equilibrium constraints, yields

$$K_1 \, dA_j/dt = K_2 \, E \, [k_{2,j-1} \, A_{j-1}A - k_{2,j} \, A_jA] + K_2 \sum_{n=1}^{j-1} k_{2,n} \, A_nA \, P_{j-n}/2$$
$$- k_{2,j} \, K_2 \, A_jA \, P_{TOT} \tag{12}$$

Equations 6, 7 and 10 may be solved simultaneously, yielding

$$A_j = k_{2,j} \, K_2 \, [E + P_{TOT}] \, A_jA / [ k_{1,j} \, K_1 \, A ] \tag{13}$$

The initial formulation of Epon 828 and NMA must be such that oxirane equivalents equal the concentration of anhydride groups. A balanced stoichiometric ratio enables the polymerization to develop macromolecules at high extents of reaction (<u>10</u>). Therefore

$$A_j = k_{2,j} \, K_2 \, A_jA / k_{1,j} \, K_1 \tag{14}$$

The concentration of carboxylic acid sites is proportional to the concentration of hydroxyl sites. Equations 9 and 14 may be solved, yielding

$$P_j = [k_{1,j} K_1 + k_{2,j} K_2] A_j A / k_{1,j} K_1 \tag{15}$$

Equation 12 may now be expressed in terms of the experimentally measurable population density distribution $P_j$.

$$dP_j/dt = k_{1,j} E \sum_{j-1} (k_{2,j-1}/k_{2,j} P_{j-1} - P_j)$$
$$+ k_{1,j} \sum_{n-1} k_{2,n} P_n P_{j-n} / 2k_{2,j} - k_{1,j} P_{TOT} P_j \tag{16}$$

Initially, oligomers of DGEBA are present in Epon 828 at a specific, but low ratio. It is also likely that some residual acid is contained within NMA (less than a few percent). Both contribute to reactive hydrogen sites. This yields the initial conditions

$$P_1(0) = A_1(0) + A_1 A(0); \quad P_j(0) = 0 \quad, j = 2,3,4... \tag{17}$$

Monomeric reactions, see Equations 2 and 3, indicate that

$$dE/dt = -C K_2 E \Sigma k_{2,j} A_j A \tag{18}$$

$$dA/dt = -C K_1 A \Sigma k_{1,j} A_j \tag{19}$$

Since the molar concentration of monomer is substantially greater than the molar concentration of polymer, Equations 14, 18 and 19 predict a balanced stoichiometry during the cure. The summation terms represent the invariant number of reactive sites on a continuously decreasing number of polymeric species.
The integration of Relationship 18 yields

$$E = E(0) \exp(-C K_2 \Sigma k_{2,j} A_j A t) \tag{20}$$

The cumulative molar concentration of polymeric species $P_{TOT}$ may be evaluated from the population density distribution, Equation 16. The first two rate expressions represent monomeric additions which do not change the molar concentration of polymeric molecules. A rate constant that describes functionality as a separable function of the molecule's degree of polymerization satisfies this constraint. The simplest, realistic function is the linear expression

$$k_{i,j} = k_{i,1} (b + m j) \tag{21}$$

When the number of molecules is reduced by a factor of two, they will on the average contain twice as many reactive sites. Summation for $j=1,2,3...$ yields

$$dP_{TOT}/dt = - P_{TOT} \Sigma k_{1,j} P_j / 2 \tag{22}$$

The sum is invariant and equals the cumulative number of reactive hydrogen sites present.
Equations 18 and 22 may be solved simultaneously, yielding the power function

$$P_{TOT} / P_{TOT}(0) = ( E / E(0) )^a \qquad (23)$$

where

$$a = \Sigma\, k_{1,j}\, P_j / (2C\, K_2\, \Sigma\, k_{2j}\, A_j A) \qquad (24)$$

Polymerization dynamics are expected to be described by Equations 16, 20, and 23.

Experimental Design

The resin contained 100.00 parts Epon 828, 80.00 parts NMA and 2.00 parts of BDMA. A part is a unit of mass. This formulation yields a resin with good mechanical performance. The formulation was cured in small test tubes that were placed in an electrically-heated, forced-air circulating oven which was controlled within $0.1°C$ of the set temperature. Specimens were removed with increasing time, thermally quenched and stored at $-25°C$.

If the content of the test tube was solid at ambient temperature, an electric drill equipped with a carbide bit produced, upon drilling, a thin ribbon that had a high surface area to volume ratio. This was leached with tetrahydrofuran at ambient conditions until chromatography analysis of the sol fraction indicated that equilibrium had been established. A gram of resin was leached with 25 ml of solvent.

A Waters Associates gel permeation (size exclusion) chromatograph was equipped with a differential refractometer and three Microstyragel columns of nominal size 100, 100 and 500 A. Chromatograms were numerically interpreted by the procedure described by Timm and co-workers ([11],[12]). A Digital LSI-11/23 microcomputer was interfaced with the chromatograph. For calibration, reactants were used for monomeric standards; linear epoxy resins were used to characterize oligomeric species leached from the resin. These polymeric standards where manufactured from phenyl glycidyl ether, nadic methyl anhydride and benzyl dimethyl amine and are distributed by a Poisson molar distribution. The numerical algorithm effectively corrects chromatograms for effects of imperfect resolution. Analyses of thermoplastic resins yield molar distributions of polymeric species that are consistent with kinetic theory ([13],[14]).

Effects of molecular configurations on hydrodynamic volume have not been explicitly incorporated into the algorithm; however, analysis of the sol fraction of thermoset resins indicates that chromatography estimates of crosslink architecture are consistent with independent observations via dynamic mechanical spectroscopy ([15-18]). Therefore, a reasonable level of confidence exists in the assignment of molecular weight in the present study. Hydrodynamic volume is a function of branching. Nonlinear chains in general have a smaller mean end-to-end distance. Zimm ([19]) calculated statistically that random branching with five trifunctional branch points per molecule reduced the radius of gyration only by a factor of about 0.7. The current research evaluates oligomeric molecules which on the average contain less than five branch points due to the relatively high ratio of bifunctional monomer to the multi-functional, but dilute polymer.

## Results

Typical data in the form of observed chromatograms are presented by Figure 1. Samples were observed for up to four hours. The oligomeric molecules within the leachate elute prior to 19.6 ml of eluent volume. For the first 1.5 hours, this material increases both in mass concentration and in average molecular weight since the peak grows in area and shifts to lower elution volumes. The oligomeric material ultimately diminished as a consequence of polymerizations with the insoluble network fraction. The monomers DGEBA and NMA elute near 21.6 and 23.6 ml, respectively. Increasing time results in diminishing concentrations.

Observed monomer concentrations are presented by Figure 2 as a function of cure time and temperature (see Equation 20). At high monomer conversions, the data appear to approach an asymptote. As the extent of network development within the resin advances, the rate of reaction diminishes. Molecular diffusion of macromolecules, initially, and of monomeric molecules, ultimately, becomes severely restricted, resulting in diffusion-controlled reactions (20). The material ultimately becomes a glass. Monomer concentration dynamics are no longer exponential decays. The rate constants become time dependent. For the cure at $60°C$, monomer concentration can be described by an exponential function.

Molecular characterization of the sol fraction produced representative data shown in Figure 3. The abscissa is the logarithm of the moles of oligomeric molecules leached per gram of resin, $P_j$. The ordinate is degree of polymerization which equals molecular weight/328 where the constant of calibration is 328. The extent of network development has a pronounced effect on oligomeric distributions as was observed with monomeric concentrations. At higher cure temperatures, i.e. higher extents of crosslink development, population density distribution data presented are representative. The locus of maxima of the population densities becomes more vertical; concentrations and average molecular weights rapidly diminish. However, oligomeric fractions at the end of the cure become dominated by exponential distributions (17), an indication that the reaction is likely more complex than that which is described by the present model.

Relationship 23 provides a method for evaluating the parameter "a" that is defined by Equation 24. The cumulative molar concentration of polymeric species $P_{TOT}$ was numerically evaluated via integration of population density distributions. The contribution of network molecules to the zeroth moment of the distribution is negligible. Results are presented by Figure 4 and show that

$$P_{TOT} = 251 \ E^{2.21} \tag{25}$$

The value for $P_{TOT}(0)$ may be evaluated since $E(0) = 1.65 \times 10^{-3}$ moles/g. Results indicate that $P_{TOT}(0)$ is about 10% of $E(0)$. This is consistent with known levels of oligomers of DGEBA in Epon 828 and the residual acid content of NMA.

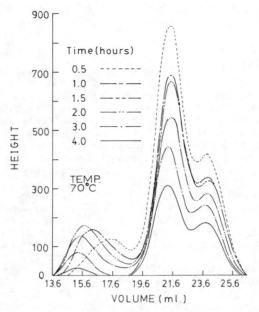

Figure 1. Chromatograms.

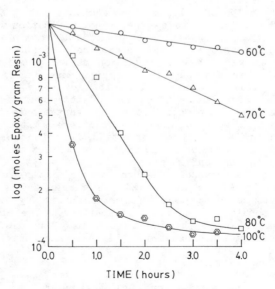

Figure 2. Monomer dynamics.

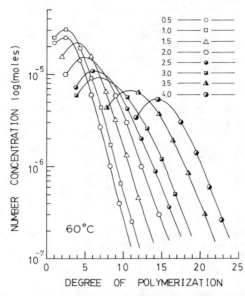

Figure 3. Population density distribution dynamics.

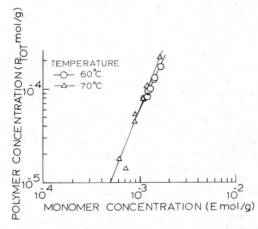

Figure 4. Correlation of monomer and polymer concentrations.

The rate constant $k_{i,j}$ is a function of reaction temperature, functionality and environmental factors which include molecular diffusion. Steric hindrance may be significant with these chemically complex monomers. These functions likely are separable. To address functionality, consider the number average molecular weight of the polymeric phase

$$MW/328 = (E(0) - E)/P_{TOT} = DP \qquad (26)$$

The number average degree of polymerization is DP. A correlation of $P_{TOT}$ as a function of the average degree of polymerization DP yielded a linear relationship, which may be generalized to

$$k_{i,j} = k_{i,1} (.15j + .85) \qquad (27)$$

Functionality, on the average, doubles when j = 8. This oligomeric specie will have an average molecular weight greater than 2500. The oligomeric species in the current study contain relatively few branch points.

Initial numerical simulations of population density dynamics incorporated experimental data of Figures 2 and 4 and Equations 16, 20, 23, and 27 into a Runge-Kutta-Gill integration algorithm (21). The constant $k_{i,1}$ was manipulated to obtain an optimum fit, both with respect to sample time and to degree of polymerization. Further modifications were necessary to improve the numerical fit of the population density distribution surface.

Chain-growth polymerizations are diffusion controlled in bulk polymerizations. This is expected to occur rapidly, even prior to network development in step-growth mechanisms. Traditionally, rate constants are expressed in terms of viscosity. In dilute solutions, viscosity is proportional to molecular weight to a power that lies between 0.6 and 0.8 (22). Melt viscosity is more complex (23). Below a critical value for the number of atoms per chain, viscosity correlates to the 1.75 power. Above this critical value, the power is nearly 3.4 for a number of thermoplastics at low shear rates. In thermosets, as the extent of conversion reaches gellation, the viscosity asymptotically increases. However, if network formation is restricted to tightly crosslinked, localized regions, viscosity may not be appreciably affected. In the current study, an exponential function of degree of polymerization was selected as a first estimate of the rate dependency on viscosity.

Studies of molecular relaxations in polymeric materials (24) clearly show the effects of cure dynamics on molecular mobility. In amorphous materials the glass transition temperature reflects long chain segmental motion. Secondary transitions involving chain ends, small chain segmental motion plus pendant group mobility are readily detected in both thermoplastics and thermosets. Small molecules significantly contribute to plasticization. If such phenomena are readily apparent with macroscopic observations that include dynamic mechanical spectroscopy and dielectric spectroscopy plus microscopic observations centered on nuclear magnetic resonance and small angle neutron scattering, one can envision that localized mobility within molecules is considerably more complex than bulk viscosity implies. Such levels of molecular activity will result in different rates of collision, which will ultimately affect rates of reaction of various sites.

The polymerization sites at the molecule's exterior are likely to experience more fruitful collisions than sites on the interior. To account for such complex phenomena, an additional separable function was introduced to simulate steric hindrance factors at the molecular level. An arbitrary, but continuous, exponential function was selected. The resulting rate constant that yields a good fit of experimental population density distributions is

$$k_{i,j} = 35 \ (.15j + .85) \ \exp(-.14j) \ \exp(-.05 \ DP) \qquad (28)$$

The first factor $k_{i,1} = 35$, is expected to be temperature dependent via an Arrhenius type relationship; the second factor defines functionality dependence on molecular size; the third factor indicates that smaller molecules are more likely to react than larger species, perhaps due to steric hindrance potentials and molecular mobility. The last term expresses a bulk diffusional effect on the inherent reactivity of all polymeric species. The specific constants were obtained by reducing a least squares objective function for the cure at $60°C$. Representative data are presented by Figure 5. The fit was good.

## Discussion

The current research shows that the model describing this step-growth polymerization is valid at relatively low conversions. Experimental monomer concentrations and the moments of the distribution are adequately fit, yielding estimates of the model parameters. The simulation demonstrates that fitting molar concentrations of polymeric species is substantially more demanding.

A kinetic simulation of a thermoset polymerization is necessarily complex. A liquid of relatively low viscosity progresses ultimately to a glassy thermoset resin. The model described approximates observed oligomeric distributions in the early stages of the cure. However, numerical errors accumulate such that the simulation becomes inadequate at higher extents of cure. The population density analysis clearly suggests that a substantial improvement in the goodness of fit can be achieved if the rate constant is allowed to be dependent on several molecular factors. Functionality increases with molecular size, but diffusion controlled rates plus steric factors diminish the reactivity of individual sites simultaneously. Future work will explore the types and forms of the separable functions necessary to achieve an optional fit at high conversions. Current exponential functions are arbitrary, but fit diverse functions for limited ranges.

A second area in addressing model improvement is in the area of experimental constraints. Multiple detectors will be integrated into the chromatogram algorithm to measure the chemical structure of eluting molecular species. One objective is to more precisely assign molecular weight for nonlinear oligomeric species. A second objective will be to measure concentrations of active sites to see if conversion affects the equilibrium distribution between alcoholic and acidic sites. A final objective will be the exploration of

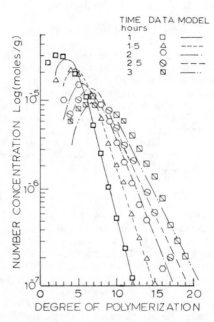

Figure 5. Numerical simulation of cure dynamics.

internal chemical links to determine the selectivity of distinct catalyst systems in yielding ester linkages as well as in developing network architecture, including crosslink average molecular weight, the average molecular weight of megamolecules and the distribution of crosslink average molecular weights.

## Acknowledgments

The authors appreciate the financial support from the Engineering Research Center and from the Brunswick Corporation, Defense Division.

## Literature Cited

1. Flory, P. J. J. Phys. Chem. 1942, 46, 132.
2. Flory, P. J. J. Amer. Chem. Soc. 1952, 74, 2718.
3. Stockmayer, W. H. J. Polym. Sci. 1953, 11, 424.
4. Stockmayer, W. H. J. Chem. Soc. 1943, 11, 45.
5. Fukui, K.; Yamabe, T. J. Polym. Sci. 1964, 2A, 3743.
6. Tanaka, Y.; Kakiuchi, H. J. Polym. Sci. 1964, 2A, 3405.
7. Fischer, R. F. J. Polym. Sci. 1960, 44, 155.
8. Morrison, R. T.; Boyd, R. N. "Organic Chemistry"; Allyn and Bacon, Inc.: Boston, MA, 1960; p. 456.
9. Shechter, L.; Wynstra, J. Ind. Eng. Chem. 1956, 48, 86.
10. Flory, P. J. "Principles of Polymer Science"; Cornell Univ. Press: Ithaca, NY, 1953; p. 92.
11. Adesanya, B. A.; Yen, H. C.; Timm, D. C. ACS Sympos. Series 1984, 245, 113.
12. Timm, D. C.; Rachow, J. W. J. Polym. Sci. 1975, 13, 1401.
13. Scamehorn, J. F.; Timm, D. C. J. Polym. Sci. 1975, 13, 1241.
14. Timm, D. C.; Kubicek, L. F. Chem. Engr. Sci. 1975, 29, 2145.
15. Ayorinde, A. J.; Lee, C. H.; Timm, D. C.; Humphrey, W. D. ACS Sympos. Series 1984, 245, 321.
16. Timm, D. C.; Ayorinde, A. J.; Huber, F. K.; Lee, C. H. Preprint Int'l. Rubber Conf. Moscow, 1984, A2, 66.
17. Timm, D. C.; Ayorinde, A. J.; Foral, R. F. British Polym. J 1984, 17, 227.
18. Timm, D. C.; Tian, W. M.; Larson, B. J.; Sudduth, R. D. J. Chromatography 1984, 316, 343-358.
19. Zimm, B. H.; Stockmayer, W. H. J. Chem. Phys. 1949, 17, 1301.
20. Gillham, J. K. British Polym. J. 1985, 17, 224.
21. Carnahan, B.; Luther, H. A.; Wilkes, J. O. "Applied Numerical Methods"; John Wiley & Sons: New York, NY, 1969; p. 363.
22. Kurata, M; Tsunashima, Y; Iwama, M; Kamada, K. In "Polymer Handbook"; Brandrup, J.; Immergut, E. H., Eds.; Wiley-Interscience: New York, NY, 1975.
23. Fox, T. G.; Gratch, S.; Loshaek, S. "Rheology-Theory and Applications"; Academic Press: New York, NY, 1956; Vol. I.
24. Nielsen, L. E., "Mechanical Properties of Polymers and Composites"; Marcel Dekker, Inc.: New York, NY, 1974; Vol. I and II.

RECEIVED December 12, 1985

# 24

# Investigation of the Self-Condensation of 2,4-Dimethylol-*o*-cresol by ¹H-NMR Spectroscopy and Computer Simulation

**Alexander P. Mgaya[1], H. James Harwood[1], and Anton Sebenik[2]**

[1] Institute of Polymer Science, University of Akron, Akron, OH 44325
[2] Kemijski Institute "Boris Kidric", Ljubljana, Yugoslavia

> The self-condensation of 2,4-dimethylol-o-cresol in pyridine solution at 100°C was studied by 60 MHz ¹H-NMR spectroscopy. Several CSMP (Continuous System Modeling Program) programs for simulating the reaction were written. One of these, when coupled with an optimization program (Chandler's STEPIT) enabled rate constants for the methylene ether forming reactions to be evaluated. The results obtained indicate that p,p-methylene ether linkages are not formed during the reaction and that the reaction occurs by nucleophilic attack of o- or p-methylol groups on o-methylol groups. This causes the condensation reaction to be much simpler than it might otherwise be expected to be.

Although the condensation of phenol with formaldehyde has been known for more than 100 years, it is only recently that the reaction could be studied in detail. Recent developments in analytical instrumentation like GC, GPC, HPLC, IR spectroscopy and NMR spectroscopy have made it possible for the intermediates involved in such reactions to be characterized and determined (1-6). In addition, high speed computers can now be used to simulate the complicated multi-component, multi-path kinetic schemes involved in phenol-formaldehyde reactions (6-27) and optimization routines can be used in conjunction with computer-based models for phenol-formaldehyde reactions to estimate, from experimental data, reaction rates for the various processes involved. The combined use of precise analytical data and of computer-based techniques to analyze such data has been very fruitful.

We previously reported a 300 MHz ¹H-NMR study on the self-condensation of 2,6-dimethylol-p-cresol (27,28) (I). When the reaction conditions are sufficiently mild, this reaction yields polymer containing only methylene ether linkages.

[Scheme showing compound I: HOCH₂-substituted phenol with CH₂OH and CH₃ groups, heating gives H₂O + [-CH₂-O-CH₂-...-]ₙ polymer]

I

There is evidence from the literature (29-33) and from our work (27,28) that this reaction occurs much more easily with o-methylol groups than with p-methylol groups. This may be attributed to intramolecular hydrogen bonding between a phenolic group and an adjacent o-methylol group, which could activate the latter toward nucleophilic displacement. Such activation would not be possible in the case of p-methylol groups. To learn more about this point, we have investigated the self-condensation of 2,4-dimethylol-o-cresol A, a compound that contains both o- and p-methylol groups.

[Reaction scheme: 2 molecules of A (2,4-dimethylol-o-cresol) react via three pathways with rate constants $k_{oo}$, $k_{op}$, and $k_{pp}$ to give products B1, B2, and B3 respectively]

## Experimental

**2,4-Dimethylol-o-cresol, A.** Thirty seven percent formalin solution (16.2g) was added to a solution of o-cresol (10g) and sodium hydroxide (4g) in 20 ml $H_2O$ at 0°C. After standing at room temperature for two days the solution was brought to pH 8.2 by the addition of 10% acetic acid and crystallization of 2,4-dimethylol-o-cresol began. The product was recrystallized from chloroform to obtain white needles melting at 93-94°C (Lit (34,35) m.p. 93-94°C). The yield was 67 percent. It is important that very pure o-cresol be used for this preparation; use of impure o-cresol impedes crystallization of the product and it remains in solution, slowly forming 2,2'-dihydroxy-3,3'-dihydroxymethyl-1,1'-dimethyldiphenylmethane II, m.p. 153-155°C (Lit m.p. 155°C (34)) in 40 percent yield.

Figure 1 shows the 60 MHz $^1$H-NMR spectrum of 2,4-dimethylol-o-cresol (2-hydroxyl-1-methyl-3,5-benzenedimethanol), A, in pyridine at room temperature.

**Kinetic Studies.** Self-condensation of A in pyridine solution was conducted at 100°C in 5 mm NMR tubes and the 60 MHz $^1$H-NMR spectra of the reaction mixtures were recorded at room temperature after various reaction times. The relative concentrations of o- and p-methylol groups and of methylene ether linkages were determined from the relative intensities of the resonances observed at $\delta = 5.25$ (o-methylol), 5.0 (p-methylol), 4.8-4.9 (methylene ether) and 2.3 ($CH_3$) ppm, respectively. Resonance areas were measured by cutting and weighing expanded spectra. Initial estimates of $k_{oo}$ were obtained by considering the reaction to be a second order process

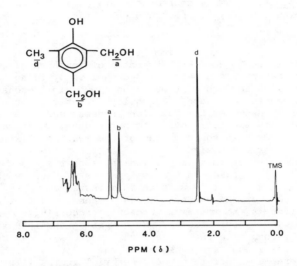

Figure 1. 60 MHz $^1$H-NMR spectrum of 2,4-dimethylol-o-cresol in pyridine at room temperature.

involving only o-methylol groups. Plots of the reciprocals of the ortho methylol group concentration (A1) versus time were constructed as shown on Figure 2 and $k_{oo}$ values were determined from the slopes of these plots. The values obtained were used as initial estimates for $k_{oo}$, $k_{op}$ and $k_{pp}$ when Chandler's STEPIT optimization program (36) was used in conjunction with a CSMP model for the reaction to obtain more appropriate $k_{oo}$, $k_{op}$ and $k_{pp}$ values.

NMR Measurements. $^1$H-NMR measurements were made at room temperature using a Varian T-60 CW-NMR spectrometer.

Results and Discussion

Figure 3 shows the 60 MHz $^1$H-NMR spectra of reaction mixtures obtained by heating 2,4-dimethylol-o-cresol in pyridine solution at 100°C for various periods of time. The o-methylol resonance at 5.25 ppm is seen to decrease in intensity, relative to the p-methylol resonance at 4.9 ppm, as the reaction proceeds and a new peak at 4.8-4.9 ppm, which is due to methylene ether groups, increases steadily in intensity. Only in the spectrum of the reaction mixture that was heated for 132 hr. is a signal due to methylene linkages evident. The chemical shift of this signal indicates that it is due to oo-type methylene linkages, based on the assignments of Hirst et al.(5). The intensity of this signal is weak in this instance. Since it is absent from most of the spectra it can be concluded that methylene ether formation is the predominate reaction occurring under the conditions employed in our studies. By comparing the intensity of the p-methylol resonance to that of o-methyl protons it was determined that p-methylol groups are consumed in the reaction, although at a rate considerably slower than the o-methylol groups. The o-methylol resonance region consists of at least two signals. One of these is due to the starting material; the other signal, which occurs at slightly higher field, is tentatively assigned to B2 and B3 dimers and to higher condensates. Several resonances are evident in the methyl resonance region, but no attempt has been made to assign them. The 60 MHz $^1$H-NMR spectra of these reaction mixtures thus provide measures of the relative concentrations of o-methylol, p-methylol and methylene ether groups as a function of reaction time. Provided that a satisfactory model is developed for the condensation reaction, it should be possible to use this information to evaluate rate constants for various reactions involved.

A Simple Model for The Reaction. Several models have been developed for the self-condensation of A. CSMP (Continuous System Modeling Program) programming has been used to integrate numerically the set of differential equations associated with each model and to calculate methylol and methylene ether group concentrations as a function of time.

The simplest model to assume is based on the possibility that the reactivities of the o- and p-methylol groups are independent of the structure of the molecule to which they are attached. According to this model, three rate constants $k_{oo}$, $k_{op}$ and $k_{pp}$ are sufficient to characterize the kinetic behavior of the reaction at a given temperature. (The subscripts associated with these rate constants define

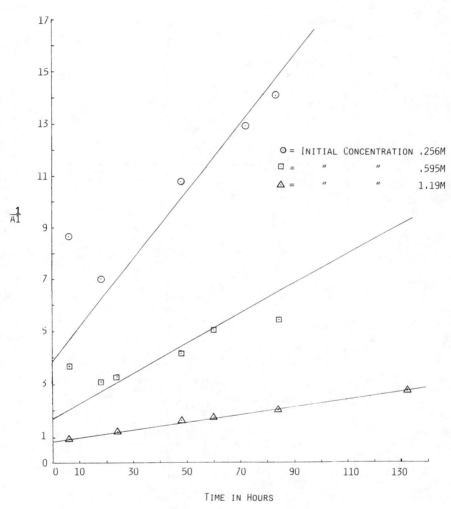

Figure 2. Plot of $\frac{1}{A1}$ versus reaction time.

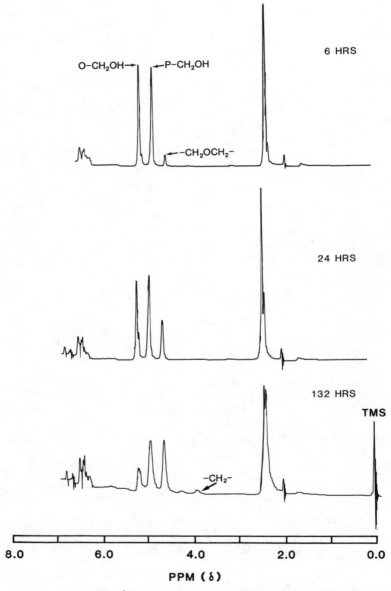

Figure 3. 60 MHz $^1$H-NMR spectra of 2,4-dimethylol-o-cresol reaction mixture in pyridine solution at 100°C.

the types of methylol groups reacting; $k_{op}$, for example denotes the rate constant for the reaction of an o-methylol group with a p-methylol group). The following differential equations describe the instantaneous rates of disappearance of o-methylol groups (A1) and p-methylol (A2) groups.

$$\frac{d(A1)}{dt} = -2k_{oo}(A1)^2 - k_{op}(A1)(A2)$$

$$\frac{d(A2)}{dt} = -2k_{pp}(A2)^2 - k_{op}(A1)(A2)$$

If the initial concentration of monomer is C, the following equations can be solved simultaneously to obtain the concentrations of o-methylol (A1), p-methylol (A2) and methylene ether (A3) groups in the reaction mixture at any time. (For simplicity, a methylene ether group is considered in this calculation to contain one carbon or two hydrogens and not two carbons or four hydrogens as the correct organic structure requires).

$$(A1) = C + \int_0^t \frac{d(A1)}{dt} = C + \int_0^t [-2k_{oo}(A1)^2 - k_{op}(A1)(A2)]dt$$

$$(A2) = C + \int_0^t \frac{d(A2)}{dt} = C + \int_0^t [-2k_{pp}(A2)^2 - k_{op}(A1)(A2)]dt$$

$$(A3) = 2C - (A1) - (A2)$$

<u>Use of CSMP in Developing the Model.</u> Solution of these equations can be accomplished by numerical methods and is easily done using CSMP programming. Program A lists a CSMP program written to accomplish this. The values of the rate constants $k_{oo}$(KOO), $k_{op}$(KOP) and $k_{pp}$ (KPP) are defined in the PARAMETER statement. The statements contained between the INITIAL and DYNAMIC statements define initial concentrations of o-methylol (ORT), p-methylol (PAR) and methylene ether (ETHER) groups. The statements that follow the DYNAMIC statement (a), define the differential equations involved in this model (DORTDT=... and DPARDT=...), (b) indicate that these are to be integrated to obtain ORT and PAR, given that the initial concentrations of these terms are equal to CONC, and (c) calculate (ETHER) as the difference between the initial concentrations of ORT + PAR and those prevailing at time t. The TIMER, FINISH and PRINT statements define the output desired and how long the calculation should occur. Figure 4 shows the output of this program, which consists of concentrations of o-methylol, p-methylol and methylene ether groups at various reaction times. Although many integration routines can be used in CSMP calculations, the variable interval Runge-Kutta method was used in this case since that is the option selected when no other method is specified.

**SIMULATION OF SELF-CONDENSATION OF 2,4-DIMETHYLOL-O-CRESOL**

| TIME | ORT | PAR | ETHER |
|---|---|---|---|
| 0.0 | 1.1900E 00 | 1.1900E 00 | 0.0 |
| 1.0000E 01 | 1.0291E 00 | 1.1624E 00 | 1.8847E-01 |
| 2.0000E 01 | 9.0430E-01 | 1.1389E 00 | 3.3680E-01 |
| 3.0000E 01 | 8.0469E-01 | 1.1185E 00 | 4.5681E-01 |
| 4.0000E 01 | 7.2339E-01 | 1.1006E 00 | 5.5605E-01 |
| 5.0000E 01 | 6.5581E-01 | 1.0846E 00 | 6.3958E-01 |
| 6.0000E 01 | 5.9876E-01 | 1.0703E 00 | 7.1094E-01 |
| 7.0000E 01 | 5.4999E-01 | 1.0574E 00 | 7.7266E-01 |
| 8.0000E 01 | 5.0781E-01 | 1.0456E 00 | 8.2661E-01 |
| 9.0000E 01 | 4.7100E-01 | 1.0348E 00 | 8.7421E-01 |
| 1.0000E 02 | 4.3860E-01 | 1.0249E 00 | 9.1654E-01 |
| 1.1000E 02 | 4.0987E-01 | 1.0157E 00 | 9.5444E-01 |
| 1.2000E 02 | 3.8422E-01 | 1.0072E 00 | 9.8859E-01 |
| 1.3000E 02 | 3.6120E-01 | 9.9926E-01 | 1.0195E 00 |
| 1.4000E 02 | 3.4041E-01 | 9.9186E-01 | 1.0477E 00 |
| 1.5000E 02 | 3.2157E-01 | 9.8493E-01 | 1.0735E 00 |
| 1.6000E 02 | 3.0441E-01 | 9.7842E-01 | 1.0972E 00 |
| 1.7000E 02 | 2.8871E-01 | 9.7229E-01 | 1.1190E 00 |
| 1.8000E 02 | 2.7431E-01 | 9.6651E-01 | 1.1392E 00 |
| 1.9000E 02 | 2.6105E-01 | 9.6104E-01 | 1.1579E 00 |
| 2.0000E 02 | 2.4881E-01 | 9.5586E-01 | 1.1753E 00 |
| 2.1000E 02 | 2.3746E-01 | 9.5095E-01 | 1.1916E 00 |
| 2.2000E 02 | 2.2693E-01 | 9.4628E-01 | 1.2068E 00 |
| 2.3000E 02 | 2.1713E-01 | 9.4184E-01 | 1.2210E 00 |
| 2.4000E 02 | 2.0798E-01 | 9.3760E-01 | 1.2344E 00 |
| 2.5000E 02 | 1.9943E-01 | 9.3356E-01 | 1.2470E 00 |
| 2.6000E 02 | 1.9142E-01 | 9.2970E-01 | 1.2589E 00 |
| 2.7000E 02 | 1.8389E-01 | 9.2601E-01 | 1.2701E 00 |
| 2.8000E 02 | 1.7682E-01 | 9.2248E-01 | 1.2807E 00 |
| 2.9000E 02 | 1.7016E-01 | 9.1909E-01 | 1.2907E 00 |
| 3.0000E 02 | 1.6388E-01 | 9.1585E-01 | 1.3003E 00 |
| 3.1000E 02 | 1.5795E-01 | 9.1273E-01 | 1.3093E 00 |
| 3.2000E 02 | 1.5233E-01 | 9.0973E-01 | 1.3179E 00 |
| 3.3000E 02 | 1.4701E-01 | 9.0685E-01 | 1.3261E 00 |
| 3.4000E 02 | 1.4197E-01 | 9.0408E-01 | 1.3340E 00 |
| 3.5000E 02 | 1.3718E-01 | 9.0141E-01 | 1.3414E 00 |
| 3.6000E 02 | 1.3262E-01 | 8.9883E-01 | 1.3485E 00 |
| 3.7000E 02 | 1.2829E-01 | 8.9635E-01 | 1.3554E 00 |
| 3.8000E 02 | 1.2416E-01 | 8.9395E-01 | 1.3619E 00 |
| 3.9000E 02 | 1.2023E-01 | 8.9161E-01 | 1.3681E 00 |
| 4.0000E 02 | 1.1647E-01 | 8.8941E-01 | 1.3741E 00 |
| 4.1000E 02 | 1.1288E-01 | 8.8725E-01 | 1.3799E 00 |
| 4.2000E 02 | 1.0945E-01 | 8.8516E-01 | 1.3854E 00 |
| 4.3000E 02 | 1.0616E-01 | 8.8314E-01 | 1.3907E 00 |
| 4.4000E 02 | 1.0302E-01 | 8.8118E-01 | 1.3958E 00 |
| 4.5000E 02 | 1.0000E-01 | 8.7929E-01 | 1.4007E 00 |
| 4.6000E 02 | 9.7109E-02 | 8.7745E-01 | 1.4054E 00 |
| 4.7000E 02 | 9.4333E-02 | 8.7568E-01 | 1.4100E 00 |
| 4.8000E 02 | 9.1668E-02 | 8.7395E-01 | 1.4144E 00 |
| 4.9000E 02 | 8.9106E-02 | 8.7228E-01 | 1.4186E 00 |
| 5.0000E 02 | 8.6643E-02 | 8.7065E-01 | 1.4227E 00 |
| 5.1000E 02 | 8.4272E-02 | 8.6908E-01 | 1.4266E 00 |
| 5.2000E 02 | 8.1990E-02 | 8.6755E-01 | 1.4305E 00 |
| 5.3000E 02 | 7.9792E-02 | 8.6606E-01 | 1.4341E 00 |
| 5.4000E 02 | 7.7673E-02 | 8.6462E-01 | 1.4377E 00 |
| 5.5000E 02 | 7.5630E-02 | 8.6321E-01 | 1.4412E 00 |

Figure 4. Output of Program A.

Use of CSMP for parameter evaluation. One could use Program A by a trial and error process to select a set of rate constants that provide the best agreement between observed o-methylol, p-methylol and methylene ether group concentrations and those calculated using the model. However this process can be automated since CSMP programs can be called as subroutines by optimization programs such as Chandler's STEPIT program (36). This enables one to evaluate parameters in CSMP programs by optimizing fits of calculated results to experimental data (26). Figure 5 shows how a combination of a CSMP program and STEPIT can be used for this purpose. Initial estimates of $k_{oo}$, $k_{op}$ and $k_{pp}$ are used in conjunction with the CSMP program to calculate methylol and methylene ether concentrations at various times. The results of these calculations are then compared with experimental results and an estimate of the goodness of fit is made. The rate

***PROGRAM A ***

```
TITLE     SIMULATION OF SELF-CONDENSATION OF 2,4-DIMETHYLOL-o-CRESOL
PARAMETER KOO=0.00544, KOP=0.00212, KPP=0.000
INCON     CONC=1.19
INITIAL
    ORT   = CONC
    PAR   = CONC
    ETHER = 0.0
DYNAMIC
    DORTDT = -2.*KOO*ORT*ORT - KOP*ORT*PAR
    DPARDT = -2.*KPP*PAR*PAR - KOP*ORT*PAR
    ORT    = INTGRL (CONC, DORTDT)
    PAR    = INTGRL (CONC, DPARDT)
    ETHER  = 2*CONC - ORT - PAR
    PRINT  ORT, PAR, ETHER
    TIMER  FINTIM=1000., DELT=1.
    FINISH ORT=3.0E-6, PAR=3.0E-6
END
STOP
ENDJOB
```

constants and error evaluation are then passed on to STEPIT which changes them and recalls the CSMP program to evaluate the suitability of the revised rate constants for generating calculated results that agree with experimental results. Under the control of STEPIT, the rate constants are systematically varied until a set of rate constants is obtained that provides the best agreement between observed and calculated results, within specified error limits.

Program B shows some of the programming that was required to accomplish this optimization. Since the modification of STEPIT for coupling with CSMP programs was discussed in an earlier publication (26), only the CSMP and BLOCK DATA subroutines required for this optimization are considered here. Both these subroutines use a labeled COMMON statement (STEPIT does also) since this is the means used to transmit most optimization information among the three programs. Considering now the CSMP subroutine, it is similar to Program

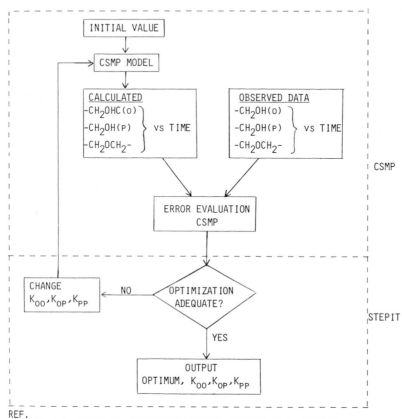

REF.
1) J.P. CHANDLER, SUBROUTINE STEPIT PROGRAM 66.1, QUANTUM CHEMISTRY PROGRAM EXCHANGE, INDIANA UNIVERSITY, BLOOMINGTON INDIANA.
2) H. JAMES HARWOOD, A. DWORAK, T.K. NYEU AND S.N. TONG, ACS SYMPOSIUM SERIES, 197, 65 (1982).

Figure 5. Evaluation of $k_{oo}$, $k_{op}$ and $k_{pp}$.

*** PROGRAM B ***

```
TITLE      SIMULATION OF SELF-CONDENSATION OF 2,4-DIMETHYLOL-o-CRESOL
/          COMMON/ONE/X(27), XMAX(27), XMIN(27), DELTAX(27),DELMIN(27),
           1 MASK(27), NV, NTRACE, MATRIX
METHOD     RKSFX
RENAME     DELMIN=DLMIN
FIXED      NV, NTRACE, MATRIX, I,J,K
INCON      CHISQ=0
INCON      CONC=1.19
INITIAL
           KOO=X(1)
           KOP=X(2)
           KPP=X(3)
           AFGEN XORT=6,.950,24,.73,48,.53,84,.42,132,.3
           AFGEN XPAR=6,.98,24,.94,48,.914,84,.882,132,.832
           AFGEN XETHER=6,.0756,24,.33,48,.529,84,.697,132,.866
DYNAMIC
DORTDT=-2*KOO*ORT*ORT - KOP*ORT*PAR
DPARDT=-2*KPP*PAR*PAR - KOP*ORT*PAR
ORT=INTGRL(CONC, DORTDT)
PAR=INTGRL(CONC, DPARDT)
ETHER=2*CONC-ORT-PAR
TIMER   FINTIM=150., DELT=1.
PROCEDURE   CHISQ=FUNCT(ORT,XORT,PAR,XPAR,ETHER,XETHER,TIME,KEEP)
           IF(KEEP.NE.1)     GO TO 1
           IF(TIME.EQ.6)     GO TO 2
           IF(TIME.EQ.24)    GO TO 2
           IF(TIME.EQ.48)    GO TO 2
           IF(TIME.EQ.84)    GO TO 2
           IF(TIME.EQ.132)   GO TO 2
           IF(TIME.EQ.134)   GO TO 3
           GO TO 1
         2 CHISQ=CHISQ+(AFGEN(XORT,TIME)*CONC-ORT)**2+(AFGEN(...
           XPAR,TIME)*CONC-PAR)**2+(AFGEN(XETHER,TIME)*CONC-ETHER)**2
         3 WRITE (6,4) KOO, KOP, KPP, CHISQ
         4 FORMAT(3F10.5,E15.5)
         1 CONTINUE
ENDPRO
TERMINAL
STOP
           BLOCK DATA
           COMMON/ONE/X,XMAX,XMIN,DELTAX,DELMIN,MASK,NV,NTRACE,MATRIX
           REAL X(27)/0.1,0.1,0.1,24*0./,XMAX(27)/1.,1.,1.,24*0./,
          1 XMIN(27)/27*0./,DELMIN(27)/0.001,0.001,0.001,24*0./,
          2 DELTAX(27)/27*0.010/,MASK(27)/27*0./
           INTEGER NV/3/,NTRACE/1/,MATRIX/0/
           END
ENDJOB
```

A except for the additional programming that is required to evaluate the agreement between calculatd and observed group concentrations (expressed by CHISQ). The METHOD statement selected a fixed step Runge-Kutta integration method rather than a variable step method because a fixed integration method is necessary for the CHISQ PROCEDURE to work properly.

The RENAME statement was necessary to avoid confusion between DELMIN variables used by the STEPIT and CSMP programs. The FIXED statement was required to define INTEGER variables defined in the COMMON statement and used in the PROCEDURE segment later in the program. The AFGEN statements are used to define a function of a variable versus TIME. In our case they defined experimental values of o-methylol group (XORT), p-methylol group (XPAR) and methylene ether group (XETHER) concentrations at various times. The values of rate constants $k_{oo}$, $k_{op}$ and $k_{pp}$ are provided to the CSMP routine by STEPIT via the X(1), X(2) and X(3) values in COMMON/ONE/. The PROCEDURE statement and the statements that fall between it and the ENDPRO statement constitute a small FORTRAN routine that is processed at each time interval to determine if the time value corresponds to any of those values for which there is experimental data, and if so to calculate the sum of the squares of the differences between observed (XORT, XPAR, XETHER) and calculated (ORT, PAR, ETHER) concentrations and to add these to CHISQ. At the end of simulation run, the value of CHISQ provides a measure of how well the rate constants are able to reproduce the observed results. This value is returned to STEPIT (via Unlabeled COMMON, see ref. 26) which then adjusts these rate constants to reduce CHISQ and thereby obtain a satisfactory fit.

The <u>BLOCK DATA</u> subroutine provides information needed for the optimization to labeled COMMON. This includes the number of variables (NV) to be optimized, their initial values (X), their maximum values (XMAX), their minimum values (XMIN), initial increments to use in varying X values (DELTAX) and an indication of how accurate the optimized variables should be (DELMIN). The parameters NTRACE and MATRIX are output options available from STEPIT. Once STEPIT has been modified for use with CSMP, it can be used without further modification. All information required for an optimization problem is provided by means of the BLOCK DATA subroutine. Although we find this approach satisfactory for batch calculations, individuals with an interactive computer system may wish to modify STEPIT so that this information can be introduced more conveniently.

<u>Results of Optimization Using Program B-STEPIT Combination</u>. Figure 6 compares o-methylol, p-methylol and methylene ether concentrations observed at various times for the self-condensation of 2,4-dimethylol-o-cresol with calculated results based on CSMP simulation (Program A) and the following rate constants

$$k_{oo} = 1.5 \times 10^{-6} M^{-1} sec^{-1}$$

$$k_{op} = 0.6 \times 10^{-6} M^{-1} sec^{-1}$$

$$k_{pp} = 0.0$$

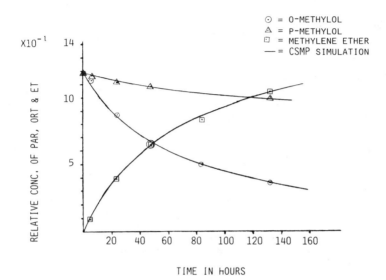

Figure 6. Changes in methylol and methylene ether group concentrations during the self-condensation of 2,4-dimethylol-o-cresol.

These rate constants were the optimum values obtained using the STEPIT-CSMP combination, starting with initial values of $k_{oo}=k_{op}=k_{pp}=27.7 \times 10^{-6} M^{-1} sec^{-1}$. The optimization required 74 trial calculations and 5.10 seconds of time on an IBM Model 158 Computer. Table I shows the path followed in the optimization. Table II shows the path of another optimization run in which $k_{oo}$ and $k_{op}$ were maintained at the values indicated above. It indicates that $k_{pp}$ is very small since a value of zero provides a better fit than any other value tried.

Figure 6 shows that CSMP calculations based on the model defined in Programs A and B yield methylol and methylene ether concentrations that are in good agreement with experimental results. The values of the rate constants evaluated by the CSMP-STEPIT combination provide interesting information about the chemistry of the condensation reaction. They show that the rate of the reaction between pairs of p-methylol groups is negligible in comparison with the rates of reaction between o-methylol groups and either o- or p-methylol groups. The fact that $k_{op}$ is approximately one-half of $k_{oo}$ is attributed to a statistical effect and it is taken as evidence that p-methylol groups can serve only as nucleophiles in these reactions whereas o-methylol groups can behave both as nucleophiles and as sites of nucleophilic displacement reactions. In the case of a reaction between a p-methylol group and an o-methylol group the reaction can apparently only occur in one way, with the p-methylol group behaving as the nucleophile and the o-methylol group as the unit activated for nucleophilic attack. However, in the case of a reaction between two o-methylol groups, either methylol group can serve as the nucleophile with the other group being the site of nucleophilic attack. This difference would cause $k_{oo}$ to be twice as large as $k_{op}$. This effect is consistent with the proposal we have discussed previously that attributes the high reactivity of o-methylol groups toward nucleophilic displacement to their ability to hydrogen bond with adjacent phenolic OH groups (27).

Given that $k_{pp}=0$ in these reactions, the self-condensation of 2,4-dimethylol-o-cresol becomes a very interesting reaction from statistical and modeling standpoints. The number of species that can be present in the system is reduced considerably by the restriction that $k_{pp}=0$. Only one o,o-methylene ether linkage can be present in each species and they must have the following general structure.

As a consequence of this simplification there are two possible dimers, two trimers, three tetramers, three pentamers, four hexamers, four heptamers, five octamers, etc. This is to be contrasted to the

Table I. Path of the Optimization Process

| Trial # | *$k_{oo}$ | *$k_{op}$ | *$k_{pp}$ | CHISQ |
|---|---|---|---|---|
| 1 | 0.110 | 0.100 | 0.100 | 18.4 |
| 2 | 0.09 | 0.100 | 0.100 | 18.1 |
| 3 | 0.07 | 0.100 | 0.100 | 17.7 |
| 4 | 0.03 | 0.100 | 0.100 | 16.1 |
| 5 | 0.03 | 0.110 | 0.100 | 16.4 |
| 6 | 0.03 | 0.09 | 0.100 | 15.9 |
| 7 | 0.03 | 0.07 | 0.100 | 15.4 |
| 8 | 0.03 | 0.03 | 0.100 | 14.1 |
| 9 | 0.03 | 0.03 | 0.110 | 14.2 |
| 10 | 0.03 | 0.03 | 0.09 | 13.9 |
| 11 | 0.03 | 0.03 | 0.07 | 13.4 |
| 12 | 0.03 | 0.03 | 0.03 | 11.6 |
| 13 | 0.02 | 0.03 | 0.03 | 10.9 |
| 14 | 0.02 | 0.02 | 0.03 | 10.1 |
| 15 | 0.02 | 0.02 | 0.02 | 9.17 |
| 16 | 0.0 | 0.0 | 0.0 | 3.97 |
| 17 | 0.01 | 0.0 | 0.0 | $1.41 \times 10^{-1}$ |
| 18 | 0.03 | 0.0 | 0.0 | $9.71 \times 10^{-1}$ |
| 19 | 0.018 | 0.0 | 0.0 | $4.69 \times 10^{-1}$ |
| 20 | 0.010 | 0.01 | 0.0 | 1.49 |
| 21 | 0.010 | 0.0 | 0.01 | 3.24 |
| 22 | 0.02 | 0.0 | 0.0 | $5.43 \times 10^{-1}$ |
| 23 | 0.0 | 0.0 | 0.0 | 3.97 |
| 24 | 0.014 | 0.0 | 0.0 | $2.71 \times 10^{-1}$ |
| 25 | 0.011 | 0.0 | 0.0 | $1.63 \times 10^{-1}$ |
| 26 | 0.009 | 0.0 | 0.0 | $1.28 \times 10^{-1}$ |
| 27 | 0.007 | 0.0 | 0.0 | $1.57 \times 10^{-1}$ |
| 28 | 0.0088 | 0.0 | 0.0 | $1.27 \times 10^{-2}$ |
| 29 | 0.0088 | 0.001 | 0.0 | $6.85 \times 10^{-1}$ |
| 30 | 0.0088 | 0.0013 | 0.001 | $2.31 \times 10^{-2}$ |
| 31 | 0.0078 | 0.0013 | 0.0 | $3.96 \times 10^{-2}$ |
| 32 | 0.0058 | 0.0013 | 0.0 | $2.91 \times 10^{-1}$ |
| 33 | 0.0018 | 0.0013 | 0.0 | $6.14 \times 10^{-2}$ |
| 34 | 0.0067 | 0.0013 | 0.0 | $2.38 \times 10^{-2}$ |
| 35 | 0.0067 | 0.0023 | 0.0 | $3.00 \times 10^{-1}$ |
| 36 | 0.0067 | 0.0003 | 0.0 | $1.17 \times 10^{-2}$ |
| 37 | 0.0067 | 0.0017 | 0.0 | $1.58 \times 10^{-1}$ |
| 38 | 0.0067 | 0.0017 | 0.001 | $1.81 \times 10^{-2}$ |
| 39 | 0;0087 | 0.0017 | 0.0 | $7.55 \times 10^{-2}$ |
| 40 | 0.0047 | 0.0017 | 0.0 | $2.93 \times 10^{-3}$ |
| 41 | 0.0061 | 0.0017 | 0.0 | $9.01 \times 10^{-2}$ |
| 42 | 0.0063 | 0.0017 | 0.0 | $1.03 \times 10^{-3}$ |
| 43 | 0.0059 | 0.0017 | 0.0 | $8.62 \times 10^{-2}$ |
| 44 | 0.0055 | 0.0017 | 0.0 | $1.09 \times 10^{-3}$ |
| 45 | 0.0059 | 0.0017 | 0.0 | $8.61 \times 10^{-3}$ |
| 46 | 0.0059 | 0.0018 | 0.0 | $7.07 \times 10^{-3}$ |
| 47 | 0.0059 | 0.0020 | 0.0 | $6.64 \times 10^{-2}$ |
| 48 | 0.0059 | 0.0024 | 0.0 | $1.55 \times 10^{-3}$ |
| 49 | 0.0059 | 0.00197 | 0.0 | $6.39 \times 10^{-3}$ |
| 50 | 0.0059 | 0.00197 | 0.001 | $7.50 \times 10$ |

Continued on next page

Table I. Continued

| Trial # | *$k_{oo}$ | *$k_{op}$ | *$k_{pp}$ | CHISQ |
|---|---|---|---|---|
| 51 | 0.0057 | 0.00197 | 0.001 | $5.51 \times 10^{-3}$ |
| 52 | 0.0053 | 0.00197 | 0.0 | $6.75 \times 10^{-3}$ |
| 53 | 0.0056 | 0.00197 | 0.0 | $5.42 \times 10^{-3}$ |
| 54 | 0.0056 | 0.00207 | 0.0 | $5.08 \times 10^{-3}$ |
| 55 | 0.0056 | 0.0023 | 0.0 | $6.94 \times 10^{-3}$ |
| 56 | 0.0056 | 0.00206 | 0.0 | $5.07 \times 10^{-3}$ |
| 57 | 0.00562 | 0.00206 | 0.0001 | $7.01 \times 10^{-3}$ |
| 58 | 0.00542 | 0.00206 | 0.0 | $5.01 \times 10^{-3}$ |
| 59 | 0.00502 | 0.00206 | 0.0 | $8.19 \times 10^{-3}$ |
| 60 | 0.0055 | 0.0026 | 0.0 | $4.91 \times 10^{-3}$ |
| 61 | 0.00529 | 0.00224 | 0.0 | $5.18 \times 10^{-3}$ |
| 62 | 0.00544 | 0.00212 | 0.0 | $4.81 \times 10^{-3}$ |
| 63 | 0.00524 | 0.00212 | 0.0 | $5.36 \times 10^{-3}$ |
| 64 | 0.00564 | 0.00212 | 0.0 | $5.29 \times 10^{-3}$ |
| 65 | 0.00544 | 0.00212 | 0.0 | $4.81 \times 10^{-3}$ |
| 66 | 0.00544 | 0.00222 | 0.0 | $5.25 \times 10^{-3}$ |
| 67 | 0.00544 | 0.00202 | 0.0 | $5.24 \times 10^{-3}$ |
| 68 | 0.00544 | 0.00212 | 0.0 | $4.81 \times 10^{-3}$ |
| 69 | 0.00544 | 0.00212 | 0.0001 | $7.30 \times 10^{-3}$ |
| 70 | 0.00544 | 0.00212 | 0.0 | $4.81 \times 10^{-3}$ |

Final values

$k_{oo} = 5.44 \times 10^{-3} \, M^{-1} hr^{-1}$

$k_{op} = 2.12 \times 10^{-3} \, M^{-1} hr^{-1}$

$k_{pp} = 0.0$

CHISQ = $4.810467 \times 10^{-3}$

* Units in $M^{-1} hr^{-1}$

Table II. Path of Optimization Process to Evaluate $k_{pp}$ Keeping $k_{oo}$ and $k_{op}$ Constant

| Trial # | * $k_{pp}$ | CHISQ |
|---|---|---|
| 1 | 0.1005 | 8.19 |
| 2 | 0.0995 | 8.17 |
| 3 | 0.0985 | 8.15 |
| 4 | 0.0965 | 8.11 |
| 5 | 0.0925 | 8.03 |
| 6. | 0.0845 | 7.85 |
| 7 | 0.0685 | 7.41 |
| 8 | 0.0365 | 5.99 |
| 9 | 0.0360 | 5.96 |
| 10 | 0.0350 | 5.89 |
| 11 | 0.0330 | 5.75 |
| 12 | 0.0290 | 5.44 |
| 13 | 0.0210 | 4.64 |
| 14 | 0.0050 | 1.55 |
| 15 | 0.0000 | $4.81 \times 10^{-3}$ |
| 16 | 0.0010 | $1.79 \times 10^{-1}$ |
| 17 | 0.0001 | $7.31 \times 10^{-3}$ |
| 18 | 0.00001 | $4.84 \times 10^{-3}$ |
| 19 | 0.00000 | $4.81 \times 10^{-3}$ |

* Units in $M^{-1} hr^{-1}$

number of species that would be present if $k_{pp} \neq 0$. Then there would be three dimers, four trimers, ten tetramers and 32 possible pentamers. We are presently developing a CSMP program that will enable us to calculate individual species concentrations as a function of time for the case that $k_{pp}=0$, since it appears that HPLC and high field NMR spectroscopy can be used to determine some of these concentrations experimentally.

Development of a Model that Involves Individual Species. As part of our efforts to develop a CSMP program for calculating the concentrations of species present in 2,4-dimethylol-o-cresol polycondensation reactions, we have written a program that calculates the concentrations of monomer and dimers during initial stages of the reaction where other species are present in negligible concentration. This program provides for the possibility that $k_{pp} \neq 0$.

In developing this program it was necessary to write expressions for the instantaneous changes in concentrations of monomer (A) and dimers (B1, B2 and B3). In the case of monomer, it was only necessary to consider reactions consuming this species, since the reaction was treated as being irreversible. The following examples show typical reactions and corresponding rate expressions.

| Reaction Involving A | Rate Term for Disappearance of A |
|---|---|
| A + A → B1(pp) | $2k_{oo}(A)(A)$ |
| A + A → B2(op) | $4k_{op}(A)(A)$ |
| A + A → B3(oo) | $2k_{pp}(A)(A)$ |
| A + B1(pp) → C1(op) | $2k_{pp}(A)(B1)$ |
| A + B1(pp) → C2(pp) | $2k_{op}(A)(B1)$ |
| A + B2(op) → C3(oo) | $2k_{pp}(A)(B2)$ |
| ⋮ | ⋮ |
| etc. | |

In these expressions, a letter is used to indicate the size of the species (A=monomer, B=dimer, C=trimer, etc.), a number is used to characterize the particular species of a given size and the letters in parentheses indicate the residual methylol functionality of the species. The coefficients for the rate terms were determined by the following three considerations:

1. The presence of two o- or p-methylol groups on a reactant requires a factor of two because the concentration of functional groups of that type is twice what it is for reactants with both o- and p-methylol groups.

2. Reactions between two identical reactants require a factor of two because both species will be consumed by the reaction, eg.

| A + A → B1 | $2k_{oo}(A)(A)$ |
| B3(oo) + B3(oo) | $2 \times 2 \times 2 \; k_{oo}(B3)(B3)$ |

3. Reactions between two reactants, each of which has one o-methylol and one p-methylol group require a factor of two because there are two ways the reaction can take place.

Rate expressions for several reactions involving B3(oo) are given below to further illustrate these concepts.

| Reaction | Rate Expressions for Disappearance of B3 |
|---|---|
| B3(oo) + A → C1(op) | $2k_{oo}(A)(B3)$ |
| B3(oo) + A → C2(oo) | $2k_{op}(A)(B3)$ |
| B3(oo) + B3(oo) → D1 | $8k_{pp}(B3)(B3)$ |
| B3(oo) + B1(pp) → D4(op) | $4k_{op}(B1)(B3)$ |
| etc. | |

Using reactions and rate expressions such as those written above, differential equations expressing the changes in monomer and dimer concentrations with time were written as follows:

$$\frac{d(A)}{dt} = -2k_{oo}(A)^2 - 2k_{pp}(A)^2 - 2k_{op}(A)^2 - k_{oo}(A)(B2) - 2k_{oo}(A)(B3)$$
$$-k_{pp}(A)(B2) - 2k_{pp}(A)(B1) - 2k_{op}(A)[(B1)+(B2)+(B3)]$$

$$\frac{d(B1)}{dt} = k_{oo}(A)^2 - 2k_{op}(A)(B1) - 2k_{pp}(A)(B1) - 8k_{pp}(B1)^2$$
$$-2k_{op}(B1)(B2) - 2k_{pp}(B1)(B2) - 4k_{op}(B1)(B3)$$

$$\frac{d(B2)}{dt} = k_{op}(A)^2 - k_{oo}(A)(B2) - 2k_{op}(A)(B2) - k_{pp}(A)(B2)$$
$$-2k_{op}(B1)(B2) - 2k_{pp}(B1)(B2) - 2k_{oo}(B2)^2$$
$$-2k_{op}(B2)^2 = 2k_{pp}(B2)^2 - 2k_{oo}(B2)(B3) - 2k_{op}(B2)(B3)$$

$$\frac{d(B3)}{dt} = k_{pp}(A)^2 - 2k_{oo}(A)(B3) - 2k_{op}(A)(B3) - 4k_{op}(B3)(B1)$$
$$-2k_{oo}(B3)(B2) - 2k_{op}(B2)(B3) - 8k_{oo}(B3)^2$$

Simultaneous integration of these equations by numerical methods can provide the concentrations of A, B1, B2 and B3 as a function of time and the concentrations of o-methylol, p-methylol and methylene ether (note earlier definition) groups can be calculated as follows, where CONC is the initial monomer concentration.

$$(ORTHO) = (A) + (B2) + 2(B3)$$
$$(PARA) = (A) + 2(B1) + (B2)$$
$$(ETHER) = 2CONC - (ORTHO) - (PARA)$$

Program C is a CSMP program based on these considerations. Figure 7 shows A,B1,B2,B3,ORTHO, PARA and ETHER concentrations calculated with the aid of this program using $k_{oo}=1.5 \times 10^{-6} M^{-1} sec^{-1}$, $k_{op}= 0.6 \times 10^{-6}, M^{-1}, sec^{-1}$ and $k_{pp}=0$. Also included in Figure 7 are experimental concentrations of o-methylol, p-methylol and methylene ether groups. Good agreement is observed between observed and calculated results for reaction times below 50 hr., indicating the general validity of our approach. We will now endeavor to generalize our programming to include species larger than monomer and dimers and will use such programming in more detailed studies of this fascinating reaction.

```
                        *** PROGRAM C ***

TITLE    SIMULATION OF SELF-CONDENSATION OF 2,4-DIMETHYLOL-o-CRESOL
PARAMETER KOO=0.00544,KOP=0.00212,KPP=0.0
INCON    CONC=1.19
INITIAL  A=CONC,B1=0.0,B2=0.0,B3=0.0,ORTHO=CONC,PARA=CONC,ETHER=0.0
DYNAMIC
DADT=-2*KOO*A*A-2*KPP*A*A-2*KOP*A*A-KOO*A*B2-2*KOO*A*B3-...
     KPP*A*B2-2*KPP*A*B1-2*KOP*A*B1-2*KOP*A*B2-2*KOP*A*B3
DB1DT=KOO*A*A-2*KOP*A*B1-2*KPP*A*B1-8*KPP*B1*B1-2*KOP*B1*B2-...
     2*KPP*B1*B2-4*KOP*B1*B3
DB2DT=KOP*A*A-KOO*A*B2-2*KOP*A*B2-KPP*A*B2-2*KOP*B1*B2-2*KPP*B1*B2-...
     2*KOO*B2*B2-2*KOP*B2*B2-2*KPP*B2*B2-2*KOO*B2*B3-2*KOP*B2*B3
DB3DT=KPP*A*A-2*KOO*A*B3-2*KOP*A*B3-4*KOP*B3*B1-2*KOO*B3*B2-...
     2*KOP*B3*B2-8*KOO*B3*B3
A=INTGRL(CONC,DADT)
B1=INTGRL(0.0,DB1DT)
B2=INTGRL(0.0,DB2DT)
B3=INTGRL(0.0,DB3DT)
ORTHO=A+B2+2*B3
PARA=A+2*B1+B2
ETHER=2*1.19-ORTHO-PARA
TIMER FINTIM=1000.,DELT=1.
FINISH A=3,0E-6,ORTHO=3.0E-6,PARA=3.0E-6
PRINT A,B1,B2,B3,ORTHO,PARA,ETHER
END
STOP
ENDJOB
```

SIMULATION OF SELF-CONDENSATION OF 2,4-DIMETHYLOL-O-CRESOL RKS INTEGRATION

| TIME | A | B1 | B2 | B3 | ORTHO | PARA | ETHER |
|---|---|---|---|---|---|---|---|
| 0.0 | 1.1900E 00 | 0.0 | 0.0 | 0.0 | 1.1900E 00 | 1.1900E 00 | 0.0 |
| 1.0000E 01 | 1.0060E 00 | 6.3629E-02 | 2.4013E-02 | 0.0 | 1.0300E 00 | 1.1572E 00 | 1.9278E-01 |
| 2.0000E 01 | 8.6759E-01 | 1.0760E-01 | 3.9351E-02 | 0.0 | 9.0694E-01 | 1.1221E 00 | 3.5091E-01 |
| 3.0000E 01 | 7.6007E-01 | 1.3900E-01 | 4.9298E-02 | 0.0 | 8.0936E-01 | 1.0874E 00 | 4.8327E-01 |
| 4.0000E 01 | 6.7431E-01 | 1.6198E-01 | 5.5758E-02 | 0.0 | 7.3007E-01 | 1.0540E 00 | 5.9591E-01 |
| 5.0000E 01 | 6.0445E-01 | 1.7912E-01 | 5.9893E-02 | 0.0 | 6.6435E-01 | 1.0226E 00 | 6.9307E-01 |
| 6.0000E 01 | 5.4655E-01 | 1.9209E-01 | 6.2441E-02 | 0.0 | 6.0899E-01 | 9.9316E-01 | 7.7784E-01 |
| 7.0000E 01 | 4.9784E-01 | 2.0200E-01 | 6.3886E-02 | 0.0 | 5.6173E-01 | 9.6574E-01 | 8.5253E-01 |
| 8.0000E 01 | 4.5636E-01 | 2.0964E-01 | 6.4553E-02 | 0.0 | 5.2091E-01 | 9.4020E-01 | 9.1889E-01 |
| 9.0000E 01 | 4.2063E-01 | 2.1556E-01 | 6.4667E-02 | 0.0 | 4.8530E-01 | 9.1641E-01 | 9.7829E-01 |
| 1.0000E 02 | 3.8958E-01 | 2.2014E-01 | 6.4386E-02 | 0.0 | 4.5397E-01 | 8.9424E-01 | 1.0318E 00 |
| 1.1000E 02 | 3.6236E-01 | 2.2368E-01 | 6.3822E-02 | 0.0 | 4.2619E-01 | 8.7354E-01 | 1.0803E 00 |
| 1.2000E 02 | 3.3833E-01 | 2.2640E-01 | 6.3056E-02 | 0.0 | 4.0139E-01 | 8.5419E-01 | 1.1244E 00 |
| 1.3000E 02 | 3.1697E-01 | 2.2847E-01 | 6.2149E-02 | 0.0 | 3.7912E-01 | 8.3607E-01 | 1.1648E 00 |
| 1.4000E 02 | 2.9787E-01 | 2.3002E-01 | 6.1143E-02 | 0.0 | 3.5902E-01 | 8.1906E-01 | 1.2019E 00 |
| 1.5000E 02 | 2.8070E-01 | 2.3115E-01 | 6.0071E-02 | 0.0 | 3.4078E-01 | 8.0308E-01 | 1.2361E 00 |
| 1.6000E 02 | 2.6520E-01 | 2.3194E-01 | 5.8958E-02 | 0.0 | 3.2415E-01 | 7.8804E-01 | 1.2678E 00 |
| 1.7000E 02 | 2.5113E-01 | 2.3245E-01 | 5.7821E-02 | 0.0 | 3.0895E-01 | 7.7384E-01 | 1.2972E 00 |
| 1.8000E 02 | 2.3831E-01 | 2.3273E-01 | 5.6674E-02 | 0.0 | 2.9498E-01 | 7.6044E-01 | 1.3246E 00 |
| 1.9000E 02 | 2.2659E-01 | 2.3282E-01 | 5.5527E-02 | 0.0 | 2.8211E-01 | 7.4775E-01 | 1.3501E 00 |
| 2.0000E 02 | 2.1583E-01 | 2.3275E-01 | 5.4388E-02 | 0.0 | 2.7022E-01 | 7.3573E-01 | 1.3740E 00 |
| 2.1000E 02 | 2.0593E-01 | 2.3256E-01 | 5.3263E-02 | 0.0 | 2.5920E-01 | 7.2432E-01 | 1.3965E 00 |
| 2.2000E 02 | 1.9679E-01 | 2.3226E-01 | 5.2156E-02 | 0.0 | 2.4895E-01 | 7.1348E-01 | 1.4176E 00 |
| 2.3000E 02 | 1.8833E-01 | 2.3188E-01 | 5.1069E-02 | 0.0 | 2.3940E-01 | 7.0316E-01 | 1.4374E 00 |
| 2.4000E 02 | 1.8048E-01 | 2.3142E-01 | 5.0005E-02 | 0.0 | 2.3049E-01 | 6.9332E-01 | 1.4562E 00 |
| 2.5000E 02 | 1.7318E-01 | 2.3090E-01 | 4.8966E-02 | 0.0 | 2.2214E-01 | 6.8394E-01 | 1.4739E 00 |
| 2.6000E 02 | 1.6636E-01 | 2.3034E-01 | 4.7952E-02 | 0.0 | 2.1431E-01 | 6.7498E-01 | 1.4907E 00 |
| 2.7000E 02 | 1.6000E-01 | 2.2973E-01 | 4.6963E-02 | 0.0 | 2.0696E-01 | 6.6642E-01 | 1.5066E 00 |
| 2.8000E 02 | 1.5404E-01 | 2.2909E-01 | 4.6000E-02 | 0.0 | 2.0004E-01 | 6.5822E-01 | 1.5217E 00 |
| 2.9000E 02 | 1.4845E-01 | 2.2843E-01 | 4.5064E-02 | 0.0 | 1.9351E-01 | 6.5036E-01 | 1.5361E 00 |
| 3.0000E 02 | 1.4319E-01 | 2.2774E-01 | 4.4153E-02 | 0.0 | 1.8735E-01 | 6.4283E-01 | 1.5498E 00 |
| 3.1000E 02 | 1.3825E-01 | 2.2704E-01 | 4.3267E-02 | 0.0 | 1.8151E-01 | 6.3560E-01 | 1.5629E 00 |
| 3.2000E 02 | 1.3359E-01 | 2.2633E-01 | 4.2406E-02 | 0.0 | 1.7599E-01 | 6.2865E-01 | 1.5754E 00 |
| 3.3000E 02 | 1.2918E-01 | 2.2561E-01 | 4.1569E-02 | 0.0 | 1.7075E-01 | 6.2196E-01 | 1.5873E 00 |
| 3.4000E 02 | 1.2502E-01 | 2.2488E-01 | 4.0756E-02 | 0.0 | 1.6578E-01 | 6.1553E-01 | 1.5987E 00 |
| 3.5000E 02 | 1.2108E-01 | 2.2414E-01 | 3.9965E-02 | 0.0 | 1.6105E-01 | 6.0934E-01 | 1.6096E 00 |
| 3.6000E 02 | 1.1734E-01 | 2.2341E-01 | 3.9197E-02 | 0.0 | 1.5654E-01 | 6.0336E-01 | 1.6201E 00 |
| 3.7000E 02 | 1.1380E-01 | 2.2268E-01 | 3.8451E-02 | 0.0 | 1.5225E-01 | 5.9761E-01 | 1.6301E 00 |
| 3.8000E 02 | 1.1043E-01 | 2.2195E-01 | 3.7726E-02 | 0.0 | 1.4815E-01 | 5.9205E-01 | 1.6398E 00 |
| 3.9000E 02 | 1.0722E-01 | 2.2122E-01 | 3.7021E-02 | 0.0 | 1.4424E-01 | 5.8668E-01 | 1.6491E 00 |
| 4.0000E 02 | 1.0417E-01 | 2.2049E-01 | 3.6336E-02 | 0.0 | 1.4050E-01 | 5.8149E-01 | 1.6580E 00 |
| 4.1000E 02 | 1.0126E-01 | 2.1977E-01 | 3.5670E-02 | 0.0 | 1.3693E-01 | 5.7647E-01 | 1.6666E 00 |
| 4.2000E 02 | 9.8478E-02 | 2.1905E-01 | 3.5022E-02 | 0.0 | 1.3350E-01 | 5.7161E-01 | 1.6749E 00 |
| 4.3000E 02 | 9.5824E-02 | 2.1835E-01 | 3.4391E-02 | 0.0 | 1.3022E-01 | 5.6691E-01 | 1.6829E 00 |
| 4.4000E 02 | 9.3286E-02 | 2.1764E-01 | 3.3778E-02 | 0.0 | 1.2706E-01 | 5.6235E-01 | 1.6906E 00 |
| 4.5000E 02 | 9.0858E-02 | 2.1695E-01 | 3.3182E-02 | 0.0 | 1.2404E-01 | 5.5793E-01 | 1.6980E 00 |
| 4.6000E 02 | 8.8533E-02 | 2.1626E-01 | 3.2601E-02 | 0.0 | 1.2113E-01 | 5.5365E-01 | 1.7052E 00 |
| 4.7000E 02 | 8.6304E-02 | 2.1558E-01 | 3.2036E-02 | 0.0 | 1.1834E-01 | 5.4950E-01 | 1.7122E 00 |
| 4.8000E 02 | 8.4165E-02 | 2.1491E-01 | 3.1485E-02 | 0.0 | 1.1565E-01 | 5.4546E-01 | 1.7189E 00 |
| 4.9000E 02 | 8.2113E-02 | 2.1424E-01 | 3.0949E-02 | 0.0 | 1.1306E-01 | 5.4155E-01 | 1.7254E 00 |
| 5.0000E 02 | 8.0141E-02 | 2.1359E-01 | 3.0427E-02 | 0.0 | 1.1057E-01 | 5.3774E-01 | 1.7317E 00 |
| 5.1000E 02 | 7.8246E-02 | 2.1294E-01 | 2.9918E-02 | 0.0 | 1.0816E-01 | 5.3404E-01 | 1.7378E 00 |
| 5.2000E 02 | 7.6422E-02 | 2.1230E-01 | 2.9422E-02 | 0.0 | 1.0584E-01 | 5.3044E-01 | 1.7437E 00 |
| 5.3000E 02 | 7.4667E-02 | 2.1167E-01 | 2.8939E-02 | 0.0 | 1.0361E-01 | 5.2694E-01 | 1.7495E 00 |
| 5.4000E 02 | 7.2976E-02 | 2.1105E-01 | 2.8468E-02 | 0.0 | 1.0144E-01 | 5.2353E-01 | 1.7550E 00 |

Figure 7. Typical output from PRINT statement in CSMP programs.

Conclusion

By simultaneously using NMR spectroscopy and computer simulation to study the self-condensation of 2,4-dimethylol-o-cresol, evidence has been obtained that only ortho methylol groups in this compound and its analogous higher condensates are activated toward reaction with nucleophilic reagents. As a result of this, no p,p-methylene ether linkages are formed in this reaction and it is much simpler than it might otherwise be. CSMP programming is very valuable for studies of this general nature and its utilization is relatively simple.

Acknowledgments

This work was supportd in part by a grant from the National Science Foundation (DMR-83-03739).

Literature Cited

1. Perrin, R.; Lamartine, R.; Bernard, G. Polymer Preprints, $\underline{24}$(2) 161 (1983).
2. Freeman, J. H. Anal. Chem., $\underline{24}$, 955 (1952).
3. Higginbottom, H. P.; Culbertson, H. M.; Woodbrey, J. C. Anal. Chem., $\underline{37}$, 1021 (1965).
4. Sebenik, A.; Lapanje, S. Angew. Makromol. Chem., $\underline{63}$, 139 (1977)
5. Hirst, R. C.; Grant, D. M.; Hoff, R. E.; Burke, W. J. J. Polym. Sci., A $\underline{3}$, 2091 (1965).
6. Szymanski, H. A.; Bluemle, A. J. Polym. Sci., A $\underline{3}$, 63 (1965).
7. Ishida, S.; Nakamoto, Y. Polymer Preprints, $\underline{24}$(2), 167 (1983).
8. Ishida, S.; Murase, M.; Kaneko, K. Polymer Preprints, $\underline{20}$, 486 (1979).
9. Ishida, S.; Tsutsuni, Y.; Kaneko, K. J. Polym. Sci., Polym. Chem., $\underline{19}$, 1609 (1981).
10. Kumar, A.; Gupta, S. K.; Kumar, B.; Somu, N. Polymer, $\underline{24}$(9), 1180 (1983).
11. Kumar, A.; Phukan, U. K.; Kulshreshtha, A. K.; Gupta, S. K. Polymer, $\underline{23}$, 215 (1982).
12. Pal, P. K.; Kumar, A.; Gupta, S. K. Polymer, $\underline{22}$, 1699 (1981).
13. Kumar, A.; Gupta, S. K.; Phukan, U. K. Polym. Eng. Sci., $\underline{21}$, 1218 (1981).
14. Kumar, A.; Kulshreshtha, A. K.; Gupta, S. K. Polymer, $\underline{21}$, 317 (1980).
15. Pal, P. K.; Kumar, A.; Gupta, S. K. Brit. Polym. J., $\underline{12}$, 121 (1980).
16. Ishida, S.; Murase, M.; Kaneko, K. Polym. J., $\underline{11}$, 835 (1979).
17. Williams, R. J. J.; Adabbo, H. E.; Arangeren, M. I.; Borrajo, J.; Vazquez, A. Polymer Preprints, $\underline{24}$(2), 169 (1983).
18. Bardey, W. K. F.; Schmidt, K. H. Polymer Preprints, $\underline{24}$(2), 171 (1983).
19. Walker, L. E.; Dietz, E. A., Jr.; Wolfe, R. A.; Dannels, B. F.; Sojka, S. A. Polymer Preprints, $\underline{24}$(2), 177 (1983).
20. Mackey, J. H.; Tiede, M. L.; Sojka, S. A.; Wolfe, R. A. Polymer Preprints, $\underline{24}$(2), 179 (1983).

21. Sebenik, A.; Lapanje, S. Angew. Makromol. Chem., 63, 139 (1977)
22. Borrajo, J.; Arangeren. M. I.; Williams, R. J. J. Polymer, 23, 263 (1982).
23. Frontini, P. M.; Cuadrado, T. R.; Williams, R. J. J. Polymer, 23, 267 (1982).
24. Zavitsas, A. A.; Beaulieu, R. D.; Leblanc, J. R. J. Polym. Sci., Al, 6, 2541 (1968).
25. Steffan, R. Angew. Makromol. Chem., 131, 25 (1985).
26. Harwood, H. J.; Dworak, A.; Nyeu, T. K.; Tong, S. N. A.C.S. Symposium Series, 45, 220 (1981).
27. Tong, S. N.; Park, K. Y.; Harwood, H. J. Polymer Preprints, 24(2), 196 (1983).
28. Tong, S. N.; Park, K. Y.; Harwood, H. J. J. Polym. Sci., Polym. Chem. Ed., 22, 1097 (1984).
29. Megson, N. J. L. Phenolic Resin Chemistry, Academic, New York, 1958, Chap. III.
30. Hultzsch, K. Chemie der Phenolharze, Springer-Verlag, Heidelberg, 1950, Chap. III.
31. Kammerer, H. Kunstoffe, 56, 154 (1966).
32. Kammerer, H.; Grossman, M.; Umsonst, G. Makromol. Chem., 39, 39 (1960).
33. Kammerer, H.; Muck, K. F.; Golzer, E.; Luder, H. Makromol. Chem., 138, 119 (1970).
34. Granger, F. S. Ind. Eng. Chem., 24, 443 (1932).
35. Barclay, M. G.; Buraway, A.; Thomson, G. H. J. Chem. Soc., 400 (1944).
36. Chandler, J. P. Subroutine STEPIT Program 66.1, Quantum Chemistry Program Exchange, Indiana University, Bloomington, Indiana.

RECEIVED March 10, 1986

# INDEXES

# Author Index

Albrecht-Mallinger, Robert, 179
Barnes, John D., 130,140
Bauer, David R., 256
Boyd, Richard H., 89
Carlson, Gary M., 241
Clark, E. S., 140
Cloeter, M. D., 275
Constien, V. G., 105,114
Dickens, Brian, 130
Dickie, Ray A., 256
Doherty, David C., 31
Eaton, B. W., 275
Fellin, E. L., 105,114
Garcia-Rubio, L. H., 202
Gilbert, Richard E., 187
Gill, T. T., 123
Golden, Joseph H., 6
Graves, G. G., 105
Hamielec, A. E., 219
Harwood, H. James, 288
Havriliak, Stephen, Jr., 76
Hild, David J., 187
Kah, A. F., 17
King, M. T., 105,114
Koehler, Mark E., 2,17,123
Krampe, Stephen E., 58
Lai, C. C., 275
MacGregor, J. F., 219
McCrackin, Frank L., 130
Mehta, J., 202
Mgaya, Alexander P., 288
Nathhorst, R. P., 155
Nave, M. D., 39
Niemann, T. F., 17
Penlidis, A., 219
Potenzone, Rudolph, Jr., 31
Provder, Theodore, 241
Rhodes, M. B., 155
Russell, Channing H., 23
Sebenik, Anton, 288
Timm, Delmar C., 187,275
Williams, T. R., 39
Wu, D. T., 170

# Subject Index

## A

Acrylic polymerization model
  capability, 172,173f
  description, 172
Administrative applications of data base management systems
  patent activity data base, 19
  patent disclosure tracking data base, 19t
Advantages of interfacing a viscoelastic device with a computer, 76-88
Advantages of modeling for coatings
  examinations of more options and more optimum options, 176
  insights into problems impractical to probe experimentally, 176
  means to correlate and communciate technical information, 176
  minimization of experimental work, 176
  routine application of complex theories, 175
  training tool, 176

Analysis of mixture models, established techniques, 61
Analysis of styrene suspension polymerization
  continuous models, 210-211
  efficiency, 211,212f,213
  free volume theory, 215,217
  initiator conversion vs. time, 215,216f
  initiator loadings, 211,214f
  kinetic parameters, 213,214t
  polydispersity, 210
  total radical concentration, 211
  variation of rate parameters as functions of conversion, 211,212f,213,216f
  variation of rate parameters vs. molecular weight, 215
Anhydride-cured epoxy polymerization, kinetic study, 275-287
Applications of mathematical dynamic model for emulsion polymerization processes
  batch and semibatch latex reactors, 225
  continuous reactors and reactor trains, 225

Applications of mathematical dynamic
    model for emulsion polymerization
    processes--Continued
  on-line state estimation, optimal
    sensor selection, and
    control, 225-226
Applications programs for automated
    hydraulic fracturing fluid
    evaluation
  automatic operation of
    viscometer, 119
  available programs, 118-119
  data analysis report, 121
  data-entry screen, 119,120f
  experiment setup file, 119
  manual operation of
    viscometer, 119,121
  menu screen, 119,120f
  real-time data display
    screen, 121-122f
  shut down program, 121
Architecture of the batch control
  system, programs, 180-181f
Automated method for hydraulic
    fracturing fluid evaluation
  application programs, 118-122
  computer and instrument
    interface, 118
  computer system, 116,117f,118
  example data, 121,122f
  instrument design, 116,117f
Automated rheology laboratory for
    fluid analysis
  computer control and data
    acquisition, 108-109
  computer interface with
    instruments, 106
  data analysis methods, 109-110
  data storage and retrieval, 110
  experimental setup, 108
Automotive enamels
  effect of cure on adhesion, 256-257
  physical properties, 256

B

Batch process control automated
    systems, 179-185
Behavior of silicone acrylate
    oligomers by designed experiments
  $2^{4-2}$ fractional
    factorial design, 40,42t
  $2^3$ full factorial central composite
    design, 51,55t
  correlation table, 42,46t
  final properties of optimized
    formulation, 56t
  histograms of responses from $2^{4-1}$
    fractional factorial
    design, 42,43f
  initial property targets, 40,42t

Behavior of silicone acrylate
    oligomers by designed
    experiments--Continued
  limitations, 46,50
  models for $2^{4-1}$ fractional
    factorial design, 46,47t
  models for constrained mixture
    design, 50t,52f
  models from $2^3$ full factorial
    central composite design, 55,56t
  range-finding experiment, 40
  regression analysis, 42,45f,51,52t
  response surface contour
    plot, 46-54
  three-component constrained mixture
    design, 50t,52f
  Yates analysis, 42,44f
Boundary region determination on
    tricoordinate contour plots
  constraints, 59
  example of feasible region
    determination and
    rescaling, 60,61,65f
  existence test of feasible
    region, 60
  Scheffe model, 59

C

Calibration for size-exclusion liquid
    chromatography, log hydrodynamic
    volume retention volume, 137,138f
Chain statistics of a random monomer
  estimation, 34-35
  example, 35-36f
Chain-growth polymerizations,
  viscosity, 284
Chemical modeling laboratory, 32-36
CLINFO, description, 24
Coatings, modeling, 170-177
Cole-Cole phenomenological equation,
    application to isochronal
    mechanical relaxation
    scans, 91-92,93f
Commerical testing laboratories
  economics of laboratory information
    management systems, 10
  use in laboratory management
    information systems, 9
  value of laboratory information
    management systems, 11
Complex algorithm-desirability
    function optimization
  advantages and disadvantages, 69
  description, 69
  three-component multiresponse
    optimization, 69-70,71-72f
Complex method
  advantage, 62-63
  analysis of four-component pesticide
    mixture, 63

# INDEX

Complex method--Continued
  correction factor, 63
  features, 62
  general procedure, 62
  optimization of 11-component glass
    formulation, 64,66,67f
  optimization of four-component flare
    mixture, 63
Composition of hydraulic fracturing
  fluids, 105-106
Computer-aided design, benefits, 31
Computer simulation, definition, 171
Computer-assisted polymer design
  current uses, 37
  goals, 38
  modeling at the bench, 36-37
  molecular modeling tools, 32-34
  polymer model building, 34
Computer-interfaced optical microscopy
  basic characterization of the
    foam, 158-159
  experimental materials, 158
  fluorescence application, 157
  fluorescence of polyurethane
    foams, 159
  image analysis, 156-157
  instrumentation, 156
  interferometric methods, 157-158
  microspectrophotometry, 157
  stereology, 160-166
Consecutive reactions,
  kinetics, 241-254
Consistency index,
  calculation, 109-110
Constrained, multivariate optimization
  algorithms, drawbacks, 61-62
Continuous system modeling program
  development of self-condensation
    model, 295-310
  output from print
    statement, 308,309f
  output of program A, 295,296f
  parameter evaluation, 297,298f
  program B description, 297,299
Control algorithm for a polystyrene
  reactor, 198,201
Cost analysis of laboratory
  information management systems
  cost-influencing factors, 12
  costs vs. benefits, 13
  total life-cycle cost, 12
Cost optimization for a polystyrene
  reactor, optimum combination of
  the inputs, 190-191
Crystallite orientation, X-ray
  diffraction, 140-153
Cure, oven optimization, 268-273
Cure chemistry
  applications, 242
  effect of materials, 242
  rate constants for each
    reaction, 242-243

Cure chemistry of epoxy
  polymerization, mechanism, 276-277
Cure in assembly ovens, 265-268
Cure window, measurement, 257
Cure windows, high-solids coatings vs.
  low-solids coatings, 265

D

Data analysis for tensile testing,
  STRESS program, 124-125
Data base management systems
  acessibility, 18
  administrative applications, 18,19t
  definition, 18
  ease of use, 18
  information retrieval
    applications, 19,20t
  laboratory applications, 20,21t
  personal applications, 22
  scope of applications, 18
  selection, 18
Data collection tasks for X-ray pole
  figure studies
  area scan, 146,149f
  description, 146,147t
  standard pole figure, 146,148f
  $\theta$ and $2\theta$ scan, 150,152f
  X and $\phi$ scan, 146
Data display functions for X-ray pole
  figure studies
  contour plotting
    routines, 148-149f,150
  description, 150
  x-y plot, 150-151,152f
Data handling for X-ray pole figure
  studies
  data bases, 143
  functions, 143,144f
  interface data base, 145
  output data file, 145-146
  parameters, 143,145
  pending task, 146
  run data base, 145
  spending task, 146
  updating function for the specimen
    data base, 143
Data handling menu for size-exclusion
  liquid chromatography, 134-138
Data processing for X-ray pole figure
  studies, hardware and software, 151
Data reduction functions for
  size-exclusion liquid
  chromatography
  base-line drift, 135,137
  control path, 135,136f
  problem, 135,138f
Decoupling for a polystyrene reactor
  insertion of dynamic lag for
    Q, 191,192f

Decoupling for a polystyrene
  reactor--Continued
 system responses, 191,193,194f
Design of mixed experiments
 boundary region determination on
   tricoordinate contour
   plots, 59-61
 multiresponse mixture
   optimization, 66-72
 objectives of a formulation, 61
 single-response mixture
   optimization, 61-67
Designed experiments, behavior of
   silicone acrylate oligomers, 40-56
2-Dimethylol-o-cresol
 $^1$H-NMR measurements, 292
 $^1$H-NMR spectra of reaction
   mixtures, 292,294f
 $^1$H-NMR spectrum, 290,291f
 kinetic studies, 290,292,293f
 preparation, 290
 self-condensation, 288
Disadvantages of modeling for coatings
 approximate nature of the
   results, 177
 availability of physical properties
   data and model parameters, 177
 long elasped time for
   development, 176
 requires multidisciplinary
   approach, 176

E

Effect of first-order lag in Q,
   responses of system to changes in
   S and MW, 192f,193,194-195f
Effect of pole placement for a
   polystyrene reactor
 performance function, 196
 poles and time constants, 193,196t
 responses of system to changes in S
   and MW, 198,199f
 state variable
   feedback, 193,196,197f
Effective cross-links, definition, 261
Elastically effective cross-link
   density
 vs. bake temperature, 263,264f,265
 calculation, 261-262
 calculation for arbitrary bake
   histories for arbitrary
   coatings, 261
 model verification for low-solids
   coating, 263t,265
 paint parameters, 263f
 vs. physical measures of cure, 261
Emulsion copolymerization model, 229
Emulsion polymerization models, 220-224

Emulsion polymerization reactors,
   mathematical modeling, 219-238
Euler-Romberg integration method,
   concentration determination, 244
Experimental design for kinetic study
   of epoxy polymerization
 equipment, 280
 resin, 280
Extensions of mathematical dynamic
   model for emulsion polymerization
   processes
 copolymer systems, 229,231f
 experimental recipes, 226t
 experimental results, 226,228f,230f
 other monomer systems, 226-230

F

Fracture analysis for tensile
   testing, 125

G

Global planning system for task
   automation, requirements, 4-5

H

Hardware for X-ray pole figure studies
 computer, 142
 diffracted intensity, 141-142
 four-circle goniometer, 141
 graphics printer, 142
 hardware configuration, 141,144f
 pulse-height analyze, 142
 specimen, 142
 X-ray and neutron
   diffractometers, 141
High-solids coatings, definition, 261
Hilds's performance function,
   definition, 196,198
Homopolymerization reaction
   engineering
 kinetic models, 202
 objective, 202
Hyaluronic acid
 monomers, 35
 trimer, 36f
Hydraulic fracturing, 105-106
Hydraulic fracturing fluid evaluation
 automated method, 115-122
 evolution of conventional
   method, 114-115

INDEX

Hydraulic fracturing fluid
    evaluation--Continued
    problems with conventional
        method, 115

I

Information retrieval applications of
    data base management systems
    commercial data base, 20t
    laboratory notebook tracking data
        base, 20
    research report index data
        base, 19,20t
Initiator efficiencies, 203-204
Instrument automation
    development, 3
    influence of microprocessors, 3
    laboratory information management
        systems, 4
    robotics systems, 3-4
Interface data base, description, 145
Internal rate of return,
    definition, 14
Isochronal mechanical relaxation scans
    advantages, 89-90
    effect of activation energy on width
        of relaxation process, 90-91,93f
    effect of L on width of relaxation
        process, 91,93f
    interpretation of relaxation
        processes, 100,101f,102,103f
    parameter determination, 92
    phenomenological
        description, 91-92,93f
    rationale, 90-91
Isocyanate
    concentration vs. time, 242-243
    isothermal cure curves, 246,247f
Isothermal reactor, description, 188

K

Kinetic model for cure optimization
    coatings, 257
    extent of reaction vs. bake
        time, 258
    mechanism of cross-linking of
        melamines, 258
    reaction of coatings, 258
Kinetic model for polystyrene
    reactors, 187-188,189t
Kinetic reaction modeling for epoxy
    polymerization, 277-280
Kinetics of polymerization cure
    chromatograms, 281,282f
    complexity of simulation, 285
    correlation of monomer and polymer
        concentrations, 281,283f

Kinetics of polymerization
    cure--Continued
    curing reactions, 276-277
    effects of cure dynamics on
        molecular mobility, 276
    experimental design, 280-281
    functionality, 284
    kinetic reaction modeling, 277-280
    model improvement, 285-287
    monomer dynamics, 281,282f
    numerical simulation of cure
        dynamics, 285,286f
    population density distribution
        dynamics, 281,283f
    reagents, 276

L

Laboratory applications of data base
    management systems
    tracking of analytical sample
        analysis requests, 21
    tracking of coatings test exposure
        data, 20,21t
Laboratory automation, new
    perspective, 2-5
Laboratory information management
    system
    application of financial analytical
        methods, 6
    in commercial testing
        laboratories, 9
    cost analysis, 12-13
    economic considerations, 9-10
    financial perspective, 10
    functions, 6-7,8t
    in quality assurance-quality control
        laboratories, 7,9
    in research and development
        laboratories, 7
    value assessment, 10-11
Laboratory information management
    systems, use in instrument
    automation, 4
Laboratory management, problems, 9
Linear polyethylene, relaxation
    behavior, 102
Linearized polystyrene reactor model
    matrix equations, 190
    steady-state values, 189t
Low-solids coatings, definition, 261

M

Material balances, 233-234
Mathematical dynamic model development
    for emulsion polymerization
    processes
    applications, 224-226

Mathematical dynamic model development for emulsion polymerization processes—Continued
  extensions, 226
  final mathematical model, 223
  material balances, 222
  molecular weight development, 222
  particle distribution determination, 223,224t
  particle size development, 222
  population balance approach, 222
Mathematical modeling for coatings
  acrylic polymerization model, 171-172,173
  advantages, 175-176
  definition, 171
  factors influencing successful application, 177
  solvent formulation system, 172,174-175
Mechanistic model for polymer property predicition, stages, 219
Menus for size-exclusion liquid chromatographic software
  auxiliary functions, 133
  data handling, 133
  housekeeping, 133
Microcomputers, acceptance, 170
Mixture design formulations, analysis and optimization, 58-72
Model development for free radical polymerization reactions
  decomposition reactions, 204,208
  grams of indicator bonded, 209
  moles of initiator, 209
  number of initiator fragments attached to polymer molecules, 209
Model development and validation for optimization of curing for thermoset coatings
  kinetic model, 257-264
  network structure model, 261-265
Molecular modeling
  at the bench, 36-37
  current uses, 37
  goals, 38
  polymer model building, 34,35f,36f
  polymeric systems, 31-38
  tools, 32-34
Molecular modeling tools, 32-34
Monodispersed approximation model, description, 220
Multiresponse mixture optimization
  choice of an algorithm, 66
  complex algorithm-desirability function optimization, 69-70,71-72f
  noniterative ordering, 66,68
  sequential generation, 66,68

N

Nelder-Mead simplex algorithm, application and description, 244

Net present value
  calculation for hypothetical laboratory information management system, 14,15t
  definition, 13-14
Network structure model for cure optimization
  description, 261-262
  extent of reaction, 261
Noniterative ordering technique
  advantages and disadvantages, 68
  description, 66,68

O

Optical microscopy, computer methods, 155-166
Oven optimization for curing
  cure response determination, 268-269
  cure uniformity vs. minimum heating rate constant, 269,270f,271
  cure window vs. oven productivity, 271,273f
  minimum bake time vs. minimum heating rate constant, 271,272f

P

Parameter determination for isochronal mechanical relaxation scans
  converged parameters, 98t,99f,100
  implementation, 94-95,98-100
  initial parameter estimates, 95t,96-97f
  method, 92,94
Particle-size development
  rate expression for particle volume, 238
  rate of change of polymer volume, 237-238
Patent activity data base, description, 19
Patent disclosure tracking data base
  activity file, 19t
  patent description file, 19t
Payback, definition, 13
PBUILD, description, 34
Personal applications of data base management systems, examples, 22
Polymerization kinetics, initiation reactions and modeling, 202-217
Polymerization processes, problems, 219-220
POLYREOM, application, 84
Polystyrene reactor, control by state variable techniques, 187-201
Population balance approach, 220,222

INDEX

Population balances
 discussion, 234-235
 nucleation term expression, 236
 total property balance equation, 236
PRANDOM, description, 34
Program A, list, 297
Program B
 description, 297
 list, 301
Program C
 list, 308
 model for self-condensation of
  2,4-dimethylol-o-cresol, 306-308,
  309f
Project activity monitoring system,
 description, 18
PROPHET, description, 23-24


Q

Quality assurance-quality control
 laboratories
  economics of laboratory information
   management systems, 9-10
  use of laboratory information
   management systems, 7,9
  value of laboratory information
   management systems, 11
Quantile plots, applications, 137,139


R

Research and development laboratories
 economics of laboratory information
  management systems, 9
 use of laboratory information
  management systems, 7
 value of laboratory information
  management systems, 10-11
Rate constant determination for curing
 concentration determination, 244
 contour plot for blocked
  isocyanate, 246,248f
 contour plot for constant
  values, 244,245f,246,247f
 contour plot using
  normalization, 250,252f,253,254f
 effectiveness, 244
 experimental, 243
 isocyanate absorbance vs.
  time, 250
 isothermal cure curves, 246,247f
 kinetics analysis flow chart, 244
 objective funcion for blocked
  isocyanate, 246,248f
 rate constants for blocked
  isocyanate, 246,249f

Rate constant determination for
 curing--Continued
  residual plot for blocked
   isocyanate, 246,249f,250
  simplest case, 244
  trial curve from contour plot
   valley, 250,253,254f
  trial curves, 250,251f
Reaction control language of batch
 control system
  basic commands, 182,183f,185f
  macros, 184,185-186f
  overview, 182
  parallel operation and flow
   control, 182,184
Relaxation processes
 dynamic shear response of isotactic
  polypropylenes, 100,101f
 relaxation behavior of linear
  polyethylene, 102
 relaxation strength vs.
  crystallinity, 100,102,103f
Research data management, packaged
 software, 23-30
Response contours, advantages, 60
Robotics systems, use in instrument
 automation, 3-4
RS/1
 applications, 29-30
 data analysis, 25,26f
 data management, 24
 description, 23
 ease of use, 25,27-28f
 extensibility, 29
 flexibility, 29
 graphics, 25,26f
 modeling, 25
 statistics, 25
 text, 25


S

Savitsky-Golay method of smoothing,
 advantages, 81,83f
Scientific software
 applications, 29-30
 data analysis, 25,26f
 early forms, 24
 ease of use, 24
 extensibility, 29
 flexibility, 29
 future developments, 30
 graphics, 25,26f
 modeling, 24
 statistics, 25
 text, 24
Scientific software packages,
 advances, 23-30
Self-condensation of
 2-dimethylol-o-cresol
  agreement between calculated and
   experimental results, 302

Self-condensation of 2-dimethylol-o-cresol--Continued
 effect of ortho vs. para methylol groups, 289
 model involving individual species, 306-310
 number of species in system, 302,306
 optimization using program B and STEPIT, 299,300f,301
 path to optimization process, 302,303-305t
 reaction under mild conditions, 288-289
 simple model using continuous system modeling program, 292-306
Sequential generation technique, advantages and disadvantages, 68
Silicone acrylate copolymers, applications, 39
Silicone acrylate oligomers
 behavior study by designed experimental techniques, 40-56
 synthesis, 40,41f
SIMPLEX program, use in oven optimization, 268-269
Single-response mixture optimization
 analysis of four-component pesticide mixture, 63-64,65f
 complex method, 62-63
 optimization of 11-component glass formulation, 64,66,67f
 optimization of four-component flare mixture, 63
Size-exclusion liquid chromatography, software for data collection and analysis, 130-139
Software for size-exclusion liquid chromatography
 calibration, 137
 data handling functions, 134-135,136f
 data reduction functions, 135,137,138f
 future directions, 137,139
 hardware overview, 131-138
 menus, 131,132f,133-134
 shell structure, 133
Software for X-ray pole figure studies
 access to system, 143
 data collection tasks, 146-150
 data display functions, 150-151,152f
 data handling, 143,144f,145-146
 data processing, 151
 extension to other computers, 151,153
 future directions, 153
 modules for pole figure facility, 142-143,144f
Solvent formulation system for coatings, 172,174-175

STEPIT, application, 299
Stereology
 basic theory of stereology, 160,162t
 computer print out, 162,163f,166f
 data collection and analysis, 162
 instrumentation, 160,161f
 parameter, 162t
 plots of parameters, 164,165f
STRESS program
 break point, 124-125
 fracture analysis, 125
 outlier identification, 125-126
 stress, 124
 work at break, 125
 yield strength, 124
Suspension polymerization of styrene
 data analysis, 210-217
 equations for intantaneous polymer properties, 204,206t
 literature data, 204,205t
 measured and molecular weight averages, 204,207f
 measured conversions and calculated polymerization rates, 204,207f
 model development, 204,208-210
Synthesis of silicone acrylate oligomers, experimental, 40,41f
System configuration for tensile testing, 123-124

T

Task automation
 definition, 4
 global planning system, 4-5
 vs. instrument automation, 2
 office automation tools, 4
 two sides of the laboratory, 4
Tensile tester, automated analysis system, 123-128
Tensile testing
 data analysis program, 124-125
 editing and reporting capability, 126,127f
 fracture analysis, 125
 importance, 123
 outlier identification, 125-126
 plotting capability, 126,128f
 system configuration, 123-124
Thermoset coatings, optimization of bake conditions, 256-273

V

Variance
 calculation, 78

Variance—Continued
  calculation of 95% confidence
    limits, 78,79t
  summary of $r^2$ values, 78,79t
Viscoelastic device-computer interface
  advantages, 76-77
  error estimation, 84,86
  estimates of experimental
    error, 78,79t,80f,81
  polymer studied, 77-78
  smoothing functions, 81,83f
  software package, 84
  viscoelastic measurements, 77
  viscoelastic properties of
    blends, 84,85f,87f
Viscoelastic measurements
  log real modulus vs.
    temperature, 78,80f

Viscoelastic measurements—Continued
  signal-to-noise ratio, 81,82f
  smoothing, 81,83f
Viscoelastic properties
  effect of temperature, 86,87f
  vs. polymer structure, 86
  vs. properties of pure
    components, 86
Viscoelastic properties of blends,
  function of
  temperature, 84,85f,87f

X

X-ray orientation studies,
  software, 140-153


*Production by Meg Marshall*
*Indexing by Deborah H. Steiner*
*Jacket design by Pamela Lewis*

*Elements typeset by Hot Type Ltd., Washington, DC*
*Printed and bound by Maple Press Co., York, PA*

## RECENT ACS BOOKS

"Chemistry and Function of Pectins"
Edited by Marshall Fishman and Joseph Jen
ACS Symposium Series 310; 286 pp; ISBN 0-8412-0974-X

"Fundamentals and Applications of Chemical Sensors"
Edited by Dennis Schuetzle and Robert Hammerle
ACS Symposium Series 309; 398 pp; ISBN 0-8412-0973-1

"Polymeric Reagents and Catalysts"
Edited by Warren T. Ford
ACS Symposium Series 308; 296 pp; ISBN 0-8412-0972-3

"Excited States and Reactive Intermediates:
Photochemistry, Photophysics, and Electrochemistry"
Edited by A. B. P. Lever
ACS Symposium Series 307; 288 pp; ISBN 0-8412-0971-5

"Artificial Intelligence Applications in Chemistry"
Edited by Bruce A. Hohne and Thomas Pierce
ACS Symposium Series 306; 408 pp; ISBN 0-8412-0966-9

"Organic Marine Geochemistry"
Edited by Mary L. Sohn
ACS Symposium Series 305; 440 pp; ISBN 0-8412-0965-0

"Fungicide Chemistry: Advances and Practical
Applications"
Edited by Maurice B. Green and Douglas A. Spilker
ACS Symposium Series 304; 184 pp; ISBN 0-8412-0963-4

"Petroleum-Derived Carbons"
Edited by John D. Bacha, John W. Newman and
J. L. White
ACS Symposium Series 303; 416 pp; ISBN 0-8412-0964-2

"Coulombic Interactions in Macromolecular Systems"
Edited by Adi Eisenberg and Fred E. Bailey
ACS Symposium Series 302; 272 pp; ISBN 0-8412-0960-X

"Historic Textile and Paper Materials: Conservation
and Characterization"
Edited by Howard L. Needles and S. Haig Zeronian
Advances in Chemistry Series 212; 464 pp; ISBN 0-8412-0900-6

"Multicomponent Polymer Materials"
Edited by D. R. Paul and L. H. Sperling
Advances in Chemistry Series 211; 354 pp; ISBN 0-8412-0899-9

For further information contact:
American Chemical Society, Sales Office
1155 16th Street NW, Washington, DC 20036
Telephone 800-424-6747